Rain and Resurrection

Rain and Resurrection

How the Talmud and Science Read the World

Irun R. Cohen

CRC Press
Taylor & Francis Group
Boca Raton London New York

CRC Press is an imprint of the
Taylor & Francis Group, an **informa** business

Rain and Resurrection:
How the Talmud and Science Read the World

CRC Press
Taylor & Francis Group
6000 Broken Sound Parkway NW, Suite 300
Boca Raton, FL 33487-2742

First issued in paperback 2019

© 2010 by Taylor & Francis Group, LLC
CRC Press is an imprint of Taylor & Francis Group, an Informa business

No claim to original U.S. Government works

ISBN-13: 978-1-58706-336-7 (pbk)

Visit the Taylor & Francis Web site at
http://www.taylorandfrancis.com

and the CRC Press Web site at
http://www.crcpress.com

Artwork by Eleanor Rubin.

Cover Art:
"World Map by Moroccan Cartographer al-Idrisi 1154." Snell, Melissa. Medieval History at About.com. Retrieved November 19, 2009 from the World Wide Web: http://historymedren.about.com/library/atlas/blatmapworld1154.htm. Used with permission.

Ancient mezuzah: ©iStockphoto.com/~User5389864_110. Used with license.

Library of Congress Cataloging-in-Publication Data

A C.I.P. catalog record for this book is available from the Library of Congress.

Dedication

The author dedicates this book to his parents Sara and Samuel T. Cohen, the creators of his life and his link to Jewishness.

Contents

Subject Two: Place

Place is a name of God. Place is a part of space defined by events. Place is the aspect of nature accessible to humans. We define the meaning of "holy" and discuss Talmudic texts that explore relationships between humans, nature, the supernatural, the devil and the creative urge.

Subject Three: Time

This section is divided into four parts. Language and time sets the stage by comparing the sense of time inherent in the tense structures of Hebrew and English. Time is central to Judaism. Indeed, God's personal Hebrew name designates a type of existence in time. Varieties of time defines chronological time, natural time, historical time and existential time. Texts of time shows us Judaism's interest in existential time. Holidays of time discusses the concepts of time—historical and existential— inherent in the character of the Jewish holidays.

Language and time

The tense structure of your language affects your sense of time; language creates your world.

Varieties of time

Time is variable; there are different ways to feel it and to measure it.

Texts of time

I interpret Talmudic texts related to time, existential time in particular.

Holidays of time

The Torah replaced the ever-cycling natural time of mythology with the progress of historical time ruled by One God. The Talmud proceeds to transform the historical message of the Biblical holidays into existential time. Existential time is the time of the individual.

Part III. Texts of Science

This section presents the idea that science, like the Talmud, depends on skillful interpretation of a received text. The text of science is nature, and science reads and writes this text according to paradigmatic rules. We discuss interpretation, understanding, signification, meaning, communication and data. We explore the idea of free will and compare the interpretation of meaning done by systems of cells with that done by collectives of people.

Understanding

Understanding involves metaphor, prediction and know-how. To understand a thing is to be proficient in interacting with it. Understanding causality, complexity and free will challenges both science and the Talmud.

Signification

Science uses the controlled experiment, an artificial slice of nature, to study nature as she really is. This strategy of signification is not unlike the art of the cinema.

Meaning

We define meaning as emerging from a process of interaction. We discuss how science tests, validates and up-dates the meaning of scientific information and scientific paradigms. Interpretation is essential to the evaluation of meaning. Biological systems create meaning through democratic process.

Communication

Texts are central to the scientific enterprise.

Data

Science generates data, enduring information about the facts of nature. But a fact of nature will not become accepted as part of the database unless the fact is noticed, interpreted and communicated by prepared minds.

Part IV. Postamble

This book has connected Talmudic texts with ideas related to science. In closing, we consider some agreements and some conflicts between the two systems of thought—religious and scientific.

Artwork

PART I

Preamble

This introduction tells you why I wrote the book and how the book is organized. We discuss the unspoken ideology of science: individuality (Person), intervention in nature (Place), and progress (Time).

§1 Narrator

No two people share the same mind: quite simply, the brain *structures itself* in response to the individual's life experience (the technical term is *self-organization*), and no two people, not even identical twins with the same genes, have the same experiences. So every individual builds a unique brain and, consequently, a unique mind. This uniqueness means that every thoughtful act performed by a human expresses to some degree the story of that individual's life. Which goes to say that every piece of writing is an autobiography, to some degree. Graphologists would claim that every handwritten word is an autobiography, to some degree. But I want to say that even a word processor, like the one I'm using now, produces a signature of the mind, a type of graphology, provided the printed sample is the product of a mind.

This book connects two fields of thought that are not very often connected: science and the Talmud. The book relates the fields by looking at two subjects common to both of them: a system of values and a method of interpretation—hermeneutics. The connection is personal. Although I studied medicine and specialized in pediatrics, I have been doing science as a professional immunologist for over thirty years. In contrast, my knowledge of the Talmud is private and limited to chance meanderings. As a youth, I studied Talmud formally for several years, but since then I have had only occasional contact with true Talmudic scholars or with their books.

This book has emerged from a series of seminars on Talmudic texts I have given sporadically over the years to my students and visiting scientists at the Weizmann Institute of Science.

The Weizmann Institute is devoted to research in biology, chemistry, physics and mathematics and to science teaching, and so you might wonder how I came to Talmudic texts.

Well, many of the students, fellows, and colleagues who visit the Weizmann come from different parts of the world: China, Japan, Korea, and India, as well as North and South America and most of the countries of Europe. These visiting scientists arrive without knowing much about Judaism and its holidays. Convening a lab seminar before major Jewish holidays allowed me to introduce the ideas and practices of the holiday to curious visitors. The natives too learned to take part. Our approach was comparative. What were the similarities and differences between Jewish thought and the thoughts inherent in the classical world religions that preceded Judaism or developed independently of it. The genesis of this book was my discovery of the intersections between ideas I saw in Talmud and some of the axiomatic beliefs underlying science. Thus, the Jewish holidays and the texts behind them, even if they be irrelevant to the conduct of research, can serve as a guide to the unspoken beliefs of the West that sanction research. The book originated through discussions with visiting scientists. But in truth, all of us who walk the earth and wonder what it's all about are visiting scientists. None hold tenure.

So be warned: This book is not a treatise about the philosophy or methodology either of science or of the Talmud. The book does not explore the history of ideas in either field, and does not compare the two fields with intellectual rigor or thoroughness. I have made no effort to review the professional literature. I cite some few works that have influenced my thinking or that can be read for more information. There is no scholarship here. This book rather gives expression to my impressions of both fields.

The book could be viewed as a series of essays. One set of essays interprets excerpts from the Talmud that, to my mind, illustrate values essential to Western science. I have extended these essays to include interpretations of the Jewish holiday year. A second set of essays interprets the function of interpretation in the method of science. Both sets of essays freely associate Talmudic and post-modern concepts. I may have constructed a chimera out of ill-fitting parts, but this monster of imagination is harmless. Indeed, the book has been a pleasure to write, and I hope it will be a pleasure to read.

Let us begin with the ideology of science.

§2 Ideology of science

An ideology, according to the Oxford English Dictionary (1989 edition), "is a systematic scheme of ideas...regarded as justifying actions"; an ideology is "held implicitly and maintained regardless of the course of events." In short, an ideology serves as an axiomatic justification for carrying out a particular course of action.

Science has been such a successful enterprise that most normal scientists are not inclined to waste time fretting about the ideology of their drive to understand nature. As we well know, success provides less incentive for soul-searching than does failure. (Do not infer that I personally question the achievements of science; I have come to thoughts of ideology only as retirement looms.) Anyway, the practice of science does presuppose some basic, usually unspoken assumptions that would seem to underlie the enterprise. These assumptions relate to the value of the individual in apposition to the judgments of society (person), the demystification of nature (place), and the progress of history (time). These three principles are self-evident to Western culture, and so we can say they constitute an ideology:

Person: A rational individual may question the teachings of authority.

Place: Humans, by right, can intervene in the workings of nature.

Time: Time is defined by progress.

These principles are fundamental to the values of science because their negation would negate science: Where the authority of received wisdom is absolute, no individual can legitimately discover a new truth. Where nature is sacred and off-limits to intervention, no experiment can be done ethically. Without the hope of progress, there is no point in intervention; the fields of science would lay fallow.

Although a scientist may never have devoted a thought to such matters, in practice he or she honors these axioms no less than a believer embraces a religious doctrine. Religions stress ideology; religions oblige their adherents to affirm the faith daily [see §39]. The axioms of science, in contrast, go without saying. The beliefs of science are natural to a scientist's way of life. Scientists, no less than persons in other walks of life, are married to their beliefs; they take them for granted. Unfortunately, relationships, even relationships to ideas, tend to deteriorate when taken for granted. Fortunately, one may sometimes invigorate a stale relationship by unconventionality.

§3 Talmudic probe

A conventional way to look at the values of science would be to analyze scientific texts or discuss the thoughts of philosophers of science. Here, we shall sidestep convention and discuss values of science by analyzing texts from the Talmud, a book manifestly unrelated to science. I do this because, at least for me, reading the Talmud is more interesting (and even closer to the spirit of science) than are the measured voices of the philosophers of science. The Talmud's approach to human values can be refreshingly unexpected. The Talmud explores in depth the importance of individuality (a concept of person), the right of humans to intervene in nature (a concept of place), and the reality of progress (a concept of time). These particular ideas, as we mentioned above [§2], are fundamental to Western science. So I believe the Talmud can stimulate the reader to think about them in a fresh way.

Some thinkers believe that Western culture owes much to ideas first expressed in the Bible; see, for example, Erich Auerbach, *Mimesis: The Representation of Reality in Western Literature*, Doubleday Anchor Books, Garden City, New York, 1957. The first chapter of *Mimesis*, Odysseus' Scar, compares the form of the Homeric narrative with the Biblical story of the binding of Isaac (Genesis 22: 1-19). Auerbach argues that the Bible is seminal to the West's concept of individuality. I believe that the Talmud can also illuminate Western values related to nature and to time that developed after the Biblical period.

The Talmud, moreover, has developed a concept of interpretation that has much in common with the way modern science interprets the text of nature. The Talmud, like science, explores complex ideas by analyzing vivid situations. The Talmud explores case histories; it does not preach an explicit doctrine; it does not formalize a fixed philosophy. The Talmud, instead of dispensing answers, asks leading questions. The Talmud is dynamic and concrete; the narrative is about people and their affairs.

Reading the Talmud can sharpen your thinking. Abstract principles are there, but they have to be extracted from the action through creative interpretation. The data uncovered by science are also concrete; abstract principles are there, but they have to be extracted from the data through creative interpretation. Now scientists invest much of their professional time interpreting the meaning of experimental results, their own results and those of their colleagues in the field. But many scientists, perhaps most, are not aware of how much the "facts" of science rest on interpretation, on hermeneutics. The lay public is even more oblivious to the difference between scientific fact and scientific interpretation. Our Talmudic texts will introduce us to a particular art of interpretation.

For a somewhat different approach to science and the Talmud, you can read *Rational Rabbis: Science and Talmudic Culture*, by Menachem Fisch, Indiana University Press, December 1997. For a classical description of the logic of science, see *The Logic of Scientific Discovery*, by Karl R. Popper, paperback edition, Routledge Classics, 2002. The sociology of scientific discovery has been analyzed by Thomas Kuhn in his book *The Structure of Scientific Revolutions, Second Edition*, Enlarged, University of Chicago Press, Chicago, 1970.

§4 Book organization

The book is divided into four parts:

I. **Preamble** is the introduction you are now reading.
II. **Talmudic Texts** begins with some basic information about the Talmud and then proceeds to analyze texts that illustrate the three subjects—Person, Place and Time. The concept of time includes an analysis of the yearly cycle of holidays outlined in the Talmud.
III. **Texts of Science** discusses science as a hermeneutic system for interpreting the texts of nature. We discuss interpretation, understanding, signification, meaning, communication and data.
IV. **Postamble** closes the book by considering the dialogue between science and religion.

Each of the four parts is divided into several sections. Each section is preceded by a précis in *italics* that describes its content and its place in the book. The sections are divided into subjects or ideas, each numbered consecutively from the beginning of the book as §1, §2, §3 and so forth. This numbering makes it possible to refer back or forward to other, related subjects by their § numbers [in brackets].

PART II

Talmudic Texts

We begin with a few words about texts in general and proceed to an introduction to the Talmud, to the logical structure of the Hebrew language, to Jewish texts, and to translating such texts. We then interpret Talmudic texts related to three subjects that intersect with science: Person, Place and Time.

§5 Text, signifier and signified

The word *Text* is derived from the Latin textus, a fabric or structure, and textus is derived in turn from *texere*, to weave. In its primary sense in English, a text is a piece of writing. Just as a cloth is woven out of raw fibers (wool, cotton, silk), a text is woven out of raw words (nouns, verbs, prepositions....) or ciphers (a mathematical formulation is also a text). A piece of writing, like a woven fabric, is a human artifact structured for human use. A text, like a fabric, has continuity. But unlike a fabric, a text is language woven to tell a story. And the aim of language is communication. A text challenges the reader to understand it; a text begs interpretation.

The association of a written text with a woven cloth is also present in Hebrew. A volume of the Talmud is called a *masekhet*; the primary Hebrew meaning of the word *masekhet* denotes the *warp* of the loom. The *warp* is the fixed foundation of threads through which the weaver actively loops the threads of *woof* to fashion a whole cloth. But note that the Hebrew *masekhet* obligates more than does the English *text*. *Text* connotes a whole cloth; *masekhet* is only the foundation of a whole cloth. A *masekhet* of Talmud is literally a *warp* that needs the *woof* of the reader's interpretation to fashion its meaning.

Interpreting a text requires the reader to see its significance; just what does the text mean? A text is like a sign in need of interpretation; what does the text signify? People who think about the science of signs, semiotics, have proposed that any sign is composed of the signifier and the signified. The signifier is the concrete form of the sign that you see, hear, smell or touch, and the signified is the idea expressed by the sign—the concept that the signifier imparts to you who interpret the sign. The signifier enters through your sense organs; the signified exits your mind. In the case of a simple sign such as a traffic light, the red light is a signifier whose meaning, or signified message is stop. In the case of a text, the signifier is the words forming the text; the signified is the interpretation of the message made by the interpreter. (See Daniel Chandler (1994): Semiotics for Beginners [www document] URL; http://www.aber.ac.uk/media/Documents/S4B/, accessed in December, 2001.)

Interpretation is important because sometimes you see an obvious signifier but don't know what it signifies (other than signifying your ignorance); the streets in Shanghai are filled with signs whose signifiers alone are clear to me (I see written words); fortunately my Chinese friend adds the signification. A high fever is a signifier (I am sick) that needs a doctor for its full signification (you have the flu).

Note, however, that the distinction between signifier and signified is an artifact, a mere abstraction; in reality a signifier cannot exist unless you already know, or seek to know what it signifies. How could you ever recognize a sign unless you already knew it signifies something (even your ignorance)? Quite simply, you have to know that a Chinese ideograph is a sign to see it as a sign. From this point of view, a signifier cannot enter your mind as a signifier unless you already have some idea about its significance. If the signifier and the signified are at all separable, the idea of the signified must actually precede your perception of the signifier—but such issues are beyond our present scope.

The concept of text is the essence of our book. What is science if not an attempt to read the text written by nature, and to re-write her too? An experiment or observation made by a scientist

constitutes a signifier of some basic truth about the essential nature of the world [§74]. The task of the scientist is to interpret just what truth the observation signifies—we shall expand this idea in part III, Texts of Science. Texts don't have to be formalities; your body is a text, your spouse is a text, your children are texts in need of interpretation. The book you write is a text that needs interpretation no less than does the book you read. We will discover that signification, the interpretation and the formulation of texts, is an activity shared both by science and by the Talmud. The logic of interpretation is not very different; only the signifiers and the methods used to interpret the signified concepts differ between science and the Talmud. To enter our Talmudic texts will require us to review a bit about the Talmud and its literary history.

§6 The Talmud

This is not a scholarly book about the Talmud; so you don't need any professional expertise to approach the texts I cite. You might want to know that the Talmud is a large work (some 60 volumes—*masekhtot*—containing about 2.5 million very concise words) that reports an oral tradition of discussions in houses of study that took place over a period of about 800 years in two geographical locations: the land of Israel and Babylon. Traditionally, and not by chance, the discussions recorded in the Talmud begin with the conquest of Israel by Alexander the Great in 332 BCE (Before the Common Era). Alexander's conquest of the East marked an intense exposure of post-Biblical Israel to the logic, power and beauty of Greek and Hellenistic cultures. One might view the beginning of the Talmud as the response of Israel and Judaism to the seduction of classical Greece. The Talmud continued its development during the periods of Roman rule and the establishment of Christianity and the Byzantine empire in the Land of Israel. Consequently, the Talmud also contains critiques of Roman and of Christian ideas and practices, but censorship, both internal and external, has left these critiques for the most part as veiled subtexts or hidden polemics.

The oral tradition of the Talmud was edited into a written form in two stages: first in Israel at about the year 200 CE, in the Hebrew language, and then again at about the year 500 CE in the Aramaic language mixed with Hebrew. The text of the first editing is called the Mishnah. The Mishnah is a concise summary of the statements of Jewish law and opinion made by various Rabbis over a period of about 500 years. The Mishnah can be viewed as an interpretation of the Bible that preceded it. The second editing of the Talmud interprets the Mishnah. This second editing summarizes 300 years of Rabbinic discussions of the text of the Mishnah. This editing, called the Talmud proper, was made in two places: in Israel and in the Jewish centers of study in Babylon. Babylon was then a part of the Sassanite Persian empire. Both versions of the Talmud are interpretations of the same Mishnah. The editing done in Israel is known as the Jerusalem Talmud. The editing in Babylon created the most influential version of the Talmud, the Babylonian Talmud. The texts in this book are from the Babylonian Talmud.

Thus the Talmud is composed of layers of texts, each text interpreting other texts: texts within texts. The primary text is the Bible, of which only selected verses are cited in the Talmud; the Mishnah is an interpretation of the commandments appearing in the Bible; the Talmud includes the Mishnah and interprets both the Mishnah and the Bible.

But the process of interpretation does not stop with the Talmud itself. For the next fifteen centuries and to this very day, Jews have continued to question, analyze and interpret Talmudic texts. Texts proliferate. Indeed, the format of a page of Talmud is effectively a *hypertext* (or *intertext*) that includes the Mishnah text, the Talmud text, and various commentaries written over the ages. For example, the comments of Rashi, 1040-1105 CE, are now included in a separate column that appears alongside the main text on every page of the Talmud. Other columns contain other texts and ancillary information. The fonts and font sizes differ to help you sort out the texts. For more detailed information about the text of the Talmud, you can read *The Essential Talmud*, by Adin Steinsaltz, Jason Aronson, Northvale, NJ, 1992.

When I studied Talmud as a boy, our rabbi told the class that God created humans with a fifth finger that had no other use but to keep the place in the Rashi column while the index finger kept the place in the main Talmud text.

Don't worry: we will not deal with the hypertext format of the Talmud or delve into Lamarckian or Darwinian explanations for the structure of the human hand in this book. I mention the above only to whet your curiosity about the antiquity of the hypertext and the cognitive utility of the body.

Various manuscripts of the Talmud show textual variations, but the first printed version (the Bomberg edition; 1520-23) established the generally accepted text and fixed the pagination. The canonical printing was probably made on the basis of the manuscripts that happened to be available to the printers. Hence, our present text is not a critical edition certified by scholars, but most of this text is probably quite accurate.

Each citation refers to the Talmud volume (*masekhet*; §5) by its traditional name, to the folio sheet by number, and to each of the two sides of the folio by *a* or *b*. Our first text, for example, is taken from the *masekhet* called Taanit, folio 7, page a: Taanit 7a. Occasionally I will cite texts that are not included in the Talmud itself, but which nevertheless were composed by the Rabbis of the Talmud. These citations do not bear the folio/page format. Citations from the Bible are by book, chapter: verse. Cited statements are indented and printed in a smaller font.

The Talmud is available in English translation with commentary; see various volumes of *The Talmud, the Steinsaltz Edition*, Random House.

§7 Cinematic Talmud

As a literary genre, I would say that some aspects of the Talmud might remind you of the art of cinema, especially as cinema has evolved in the second half of the 20th Century. The Talmud, like cinema, is built on dialogue. Of course, other literary genres too use dialogue, but dialogue is essential to the Talmud. The Talmud was transmitted orally before being written down; the Talmud is built of sound-bites. The Talmud uses visual shots; short action clips abound. There are abrupt shifts of scene, flashbacks, jumps into the future; panoramic long-shots alternate with close-ups; the real and the surreal change places. Many Talmudic episodes only make sense when you visualize the scene in your mind's eye. Some of the scenarios are like a shooting script—they proceed *on location*. Talmudic time, like cinematic time, can be condensed; dialogues can take place between persons who lived a thousand years apart. The personae include bit-part actors along with familiar stars who appear and reappear in varied settings. Indeed, the different roles played by a single actor in paradoxical situations invite interpretation; a certified saint can chase a woman (for example, Rabbi Akiva; §26 and §39).

Above all, the art of the Talmud, like that of the cinema, is created by *montage*. The sequence of ideas and events is flexible; there is no obviously fixed structure. The editing of associations makes the text. Montage is meaningful even in the dry Talmudic discussions that tell no story. The flow of questions is like a flow of scenes; the apposition and timing of situations is the art. Indeed, the context of subjects into which an episode is inserted is itself a commentary, an unspoken meta-discussion; what does the present discussion have to do with the preceding and subsequent discussions? Like the cinema, Talmudic action has no beginning; the action just starts. Unlike cinema, however, Talmudic narrative has no clear end; it just goes on to a new situation. (Science, too, is a story without end.) Most Talmudic discussions are not cinematic; but still I think that the Talmud at its best creates enchanting screenplay. Unless you already know the Talmud, you'll just have to read on and judge for yourself.

§8 Translating the Talmud

The translation of the Talmudic texts into English was done by me. I have used the Talmud edited by Adin Steinsaltz (Israel Institute for Talmudic Publications, Jerusalem, Israel) and have followed its Hebrew translation of the Aramaic, when my own rudimentary Aramaic failed me. A true English rendering of the Talmud is difficult, and, to be truthful, really not possible. The reasons are two: the barrier of the language and the character of the literature. Hebrew and Aramaic are closely related Semitic languages that are equally distant from English and from other European languages in semantics, syntax, rhythm and internal logic. Moreover, the written text itself is a transcription from the collective memory of an oral tradition; the Talmud was maintained in memory long before it was committed to writing. To this day, the Talmud is not read but chanted aloud; its meaning, even in the original tongues, is inherent in the melody of its words and phrases. The Talmud lacks punctuation, so commas, stops, questions, exclamations, and emphases are all created by the song. The written text is analogous to a musical score; it invites interpretation and variation; it requires practice and virtuosity. The oral tradition still lives there; it's part of the screenplay.

Biblical literature is terse, compressed, concise, brief and pithy; it begs interpretation. The Talmud is more so: succinct to the extreme, filled with epigrams and aphorisms. The Talmud, like any literature, evades translation. Yet, I have had to translate the chosen texts; otherwise there would be no book. I have avoided a word-for-word representation, which would be meaningless, and have tried to bring you the sense of the matter. Such a translation is necessarily an interpretation. There is no alternative.

§9 The Hebrew roots of interpretation and reality

Most Hebrew words are formed from roots of two, usually three and rarely four consonants that bear a general meaning, a kind of semantic value. These root consonants are poured into a limited set of linguistic forms composed of vowels and auxiliary consonants. The specific meaning of each word is thus derived from the general meaning inherent in its root consonants plus the added particulars of the vowel and auxiliary forms. Different root consonants are expressed (can only be pronounced) in the clasp of standardized vowel formats. In this way, tens of thousands of different words emerge from some two thousand roots. All languages are logical, but the logic of Hebrew is conspicuous.

The very logistics of reading and writing Hebrew foster interpretation: Since the vowel formats of Hebrew are relatively fixed, Hebrew can be written without vowel signs. Vowel signs, in fact, are absent from Hebrew texts, ancient and modern. Vowel signs were devised only in the 8th Century to preserve the proper Hebrew pronunciation of sacred texts, and they are used today only in poetry or in teaching texts, or when needed to dispel ambiguity. To read, the mature Hebrew reader inserts the vowels mentally into the text, on the fly. But how does the reader know which vowels to insert when he or she sees an otherwise unpronounceable set of consonants? A particular set of consonants could have different meanings with different sets of vowels. Some ancillary consonants are used to indicate certain vowels in Hebrew words, but for the most part, the reader's mind chooses the right vowels by reading the context of the whole phrase, sentence or subject matter. Let me furnish you with a hypothetical example from English: consider the consonants RD; how would you know whether the word should be read as ReaD, RiD, ReeD, ReD, RoDe, RaiD, RuDe, RiDe? You could do it, but you would have to know more about the context; RD instruments are never colored RD; you RD the horse; you RD the book. I'm sure you had no problem with the foregoing. A closer analogy to reading Hebrew would look like this: RD NSTRMNTS R NVR CLRD RD; Y RD TH HRS; Y RD TH BK.

Now note this: The meanings of any word in any language are colored by the context in which the word is used. But in Hebrew (and Aramaic), even the phonetics of a word depend on its context. You can say an isolated English word, but you cannot voice a naked Hebrew word without perceiving or inventing some context that suggests which vowels to use. You can read English thoughtlessly without interpreting it. Reading Hebrew, in contrast, is interpreting Hebrew. A Hebrew text must be interpreted to be read. Any Hebrew text engages the mind, even at first glance. Maybe that's why many Hebrew texts became sacred.

The room for interpretation provided by Hebrew can be better appreciated when we consider what the Greeks did to the Hebrew (Phoenician) alphabet when they imported it into Europe. The early Greek alphabet was a modification of the original Semitic alphabet used by the Hebrews and Phoenicians (ancient Phoenician was very close to Biblical Hebrew). Now this original Semitic alphabet contained the twenty-two letters present in modern Hebrew, all consonants; none of the letters were vowels. The Greek alphabet (and the Latin and other alphabets derived from the Greek alphabet) all contain vowels; where did they come from? Well, it seems that the Greeks created the vowel signs A, E and O by converting three letters that originally represented Semitic guttural consonants not present in Greek. I and U were converted into vowels from other Semitic consonants. In essence then, the Greeks made the alphabet completely phonetic by doing away with the ambiguity inherent in written Hebrew. You could then read Greek—and so can now read English—without having to interpret the context of the words, as you must do to this very day when you read Hebrew. The Greeks thus added a significant level of clarity to the Semitic alphabet, on its way to becoming a Greek alphabet. How classically Greek to endow writing and reading with transparent meaning at the phonetic level. Greek logic only had room for *true* and *false*; mathematics, for Plato, was the only true language. The Greek philosophers did not leave any corners of reality in darkness; all was illuminated. (Obviously, I have oversimplified Greek culture, but only to make a contrasting point about the Hebrew and Talmudic views of reality; Sophocles and other Greek poets, like poets in every culture, were quite familiar with the ambiguity of human language and with the dark mystery of human desires. Despite the running commentary provided by the Chorus, Greek plays do leave room for interpretation.)

But of course, there is no free lunch. In dispelling the ambiguity of the written word, the Greeks destroyed some of the mystery of context and much of the need for primary interpretation. Reading became more a technology and less an art of the reader. Auerbach, in comparing Homer with Genesis, emphasizes the unambiguous surface clarity of the Greek text compared to the mystery of the Hebrew text that invites, even demands interpretation—see Erich Auerbach, *Mimesis: The Representation of Reality in Western Literature*, cited above [§3]. Thus, the Greek quest for clarity (and distaste for dark interpretation) included even the symbolic domain of the alphabet. The Hebrew alphabet had to be clarified and completed by Greek vowel signs.

The Rabbis, by way of contrast to the Greeks, loved to unleash the literal words of the Torah for new interpretations by proposing alternative vowel readings at key points in the text. By changing the vowels, the Rabbis could change the meaning of the words; in Hebrew you can easily exchange *children* and *builders* [*banim* versus *bonim*] by fiddling with the unwritten vowels. Thus the context of a Hebrew sentence not only determines the vowels of a word, the vowels of a single Hebrew word can determine the contextual meaning of a whole Hebrew sentence. To provide an opening for an apt interpretation, the Rabbis, albeit rarely, will even switch root consonants. How un-Greek; how Talmudic.

(Actually, the English translation of the Bible contains one example of vowel-play—clearly unintentional: Exodus 10:1 and other verses in the King James version mention the Red sea, but the Hebrew original says Reed sea (*yam suf*; *suf=reed*; and not *yam adom*; *adom=red*.)

Indeed, Hebrew reality is essentially linked to spoken language. Examples abound. The Hebrew word for Scripture is *miqra*; *miqra* does not mean *writing* (*script=scripture*), but *reading*, *speaking*, *calling*. The English word *book* derives from a writing tablet; a book in English is physical. The Hebrew word for *book* is *sefer*; the root of *sefer*, S-F/P-R, denotes *telling* or *saying*, and not the physical document. Thus the essence of the Hebrew book is not to write it or to read it, but to tell it and to hear it. Hebrew understanding too is verbal: A command to *hear* is *shma*, and a command to *understand* is also *shma* [§39]. The Hebrew word for *meaning* (*mashmaut*: literally, it makes us hear) is derived from this same *shma* that denotes both hearing and understanding [§76].

(The Talmud states the importance of hearing in its own concrete way. Baba Kama, 85b, lists the payment of damages for causing the loss of various organs; the loss of an arm, leg or eye requires reimbursement for the value of an arm, a leg or an eye. But the perpetrator of deafness must pay the victim the cost of a whole person—the Talmud leaves it up to the reader to interpret why it holds the loss of hearing to be so devastating. To learn how to price a person and assorted organs, you can study the discussion in Baba Kama.)

In Hebrew, even concrete entities are verbal: The Hebrew word *davar* can mean a *word* and also a *thing*, *matter* or *object*. The root of *davar* (D-V/B-R) means to *speak*. Words, at least in Hebrew, are human reality, not mere signs of reality. Aristotle thought otherwise, but that would take us too far afield.

Let us close this detour into the linguistics of reality by examining the meanings of some actual terms. What does the word *Torah* signify? The root of the word Torah is Y-R-H. Note that the Y of the root seems to have become a T in the word Torah; the T is an auxiliary consonant. The important point for us now is that the root of Torah is related to the roots of the words for teacher (*moreh*) and for parent (*horeh*). The word Torah means a body of *teaching*. The parent and the teacher are entwined there too. The word Talmud means the act of *learning*. The three-consonant root of Talmud is L-M-D, and the word *student* (*talmid*) comes from the same root. The word Mishnah, the first level of the Talmud text [§6], has the root SH-N-H; *shoneh* means to *repeat*, and hence to *study* (by repetition), and *shanah* means the cycle of the *year* [§56]. *Shoneh* also means *different*, or that which does not cycle. In other words, *repetition* (that which cycles) and *non-repetition* (that which is different or singular) use the same root.

In fact, it is not uncommon for a Hebrew root to encode two opposite meanings. This paradox might be explained by the fact that any meaning carries the shadow of its exact negative image. For example, the Hebrew words for *strange* and *familiar* share the same root; they are the opposite poles of *relationship*. To *implant* and to *uproot* share the same root; they too are the opposite poles of a generic action. Even the root D-V/B-R, which we saw above has the constructive meaning of *word/thing*, also gives rise to the words for *destruction/plague/desert*; doing and undoing are extreme opposites: *davar = word*, a vehicle that creates worlds; *dever = plague*, a vehicle that destroys worlds (is the Hindu god Shiva, the creator/destroyer, a personification of the Hebrew root D-V-R?) Thus, a particular Hebrew root might be considered analogous to a numeral that can accommodate either a plus (+) or a minus (-) sign without changing its absolute value. Hebrew roots would seem to provide an insight into the way the mind symbolizes meaning, but that is beyond our present interest.

The term *rav* means *great* and metaphorically *master*, and so *rabbi* means *my master*; Rav and Rabbi designate the people who created the Talmudic texts. The Rabbis all were men, although women appear prominently in the text. The Rabbis undertook to mold the laws recorded in the Torah into a concrete way of life, the *Halakhah*. The *Halakhah* is Judaism's code of behavior. The *Halakhah* is not a single text; it is the evolving collective of obligatory practices shouldered by observant Jews. The root of *Halakhah* is H-L-KH, which connotes the act of walking. The *Halakhah* is a *path*, or a *way*. But the *Halakhah* is not a path passively waiting to be walked; *Halakhah* is a way of proceeding by doing. *Halakhah* is the way one is to carry out in practice the commandments prescribed by Judaism. Christianity diverged from Talmudic Judaism when Christianity departed from the *Halakhah*.

One might say that the root system of Hebrew, which highlights the semantic relationships between different words, turns every Hebrew word into a text; each word bears a residue of meaning common to all the other words that share the same root. Hence, a Hebrew word can trigger a chain of semantic associations beyond the relatively specific meaning *denoted* by the word itself. Just as the Hebrew alphabet demands interpretation, the root system of Hebrew invites *connotation*.

Indeed, the connotations of Hebrew words have been magnified by the very conservation of the language. Living Hebrew is an ancient language preserved by its sacred texts; the three thousand year-old Hebrew of the Bible lives and can be understood by even a child who speaks Hebrew; the two thousand year-old Hebrew of the Mishnah is essentially modern Hebrew. (How many educated speakers of English can understand Chaucer?)

Thus, a turn of phrase, sometimes even a single word, can call to mind a passage from the Bible, a dictum from the Talmud, a phrase from the prayer book. It is no wonder that Adam and Eve enjoyed speaking Hebrew.

§10 Levels of Jewish text

Conceptually, we can think of Torah, Talmud, and *Halakhah* as three levels of related texts: the Torah is the primal text, the fundamental text; the Talmud is the perceptual text that perceives the Torah and proposes to interpret it; the *Halakhah* is the translation of the Talmudic text into a functional program of behavior.

Let us consider a few examples of this chain of texts. The Torah states that the seventh day of the week is the Sabbath, a day of rest during which it is forbidden to work or to kindle fire. The Sabbath is a principle of Judaism, yet the Torah makes the briefest of statements about it; how is a Jew to observe the day? The *Halakhah* is the code that complements the Torah by providing the details. The *Halakhah* prescribes the way a Jew should observe the Sabbath; the fulfillment of the Sabbath includes home rituals that usher in the day like lighting candles, blessings over wine and bread, special prayers in the synagogue, regulations regarding travel, effort, and emergency situations, and so on to the ceremony that marks the end of the Sabbath day. How do the few words of the Torah text become transformed into a functional program for the Sabbath; how does one go from written word (Torah) to concrete practice (*Halakhah*)? The Talmud is the bridge. The Talmudic discussion is a process that transforms the primal Torah narrative into a prescription for living, the *Halakhah*. Through an ingenious and eclectic interpretation of Torah texts the Talmud, as we shall see, defines meaning and principles.

What, in the case of the Sabbath, is work? What is time? What is effort? What is travel? The principles developed in the Talmud form the basis of the functional prescriptions listed in the *Halakhah*. The Torah is the raw material; the Talmud is the text that explores ideas; the *Halakhah* is the functional code. Recall again that Talmud itself is constructed of two levels of text: the Mishnah and the Talmud proper [§6].

Consider other matters. The Torah mentions charity, justice, ownership, sale, courts, testimony, marriage, relationship, and many other types of obligation and responsibility. The Talmud analyzes every aspect of each issue to provide a conceptual framework for defining the poor and their rights, the rich and their obligations, the courts and their procedures, the tests of truth and evidence, relationships between man and woman, between parent and child, between teacher and student, and between employer and employee. The Talmud delves into the thoughts, desires and satisfactions of humans, rational and irrational, natural and unnatural. The Talmud connects the words of the Torah to living in this world (*Halakhah*). The Talmud analyzes the Torah within the constraints of nature, dealing with time, space, matter, life, death, society, psychology, spirit and thought. The Talmud gives rise to *Halakhah* in the way that basic science gives rise to technology.

§11 The Yeshiva

An assembly where Talmudic discussions take place is called a Yeshiva. A Yeshiva is like a parliament. The word *parliament* denotes a place where people assemble to *speak*; the Hebrew word *yeshiva* designates the act of *sitting*. Contrast sitting with walking; the Yeshiva, *sitting*, carries the connotation of pausing, of stopping to think. *Halakhah*, in contrast, is *walking*, is moving on in the world [§9].

The institution of the Yeshiva has been in practice among Jews the world over for more than two thousand years. If you visit a Yeshiva, despite its name, you will see many people standing and talking, as well sitting and thinking. The Yeshiva is a laboratory for testing Talmudic ideas through thought-experiments. A Yeshiva is also called a *Bet Midrash*, which means a *House of Interpretation*, or a *House of Questioning*.

§12 Disclaimer

I have chosen for this book Talmudic excerpts that suit my purpose. This text is personal, not scholarly. Professional Talmudic scholars may disagree with my choice of texts and with the interpretations I give them, but freedom of interpretation is guaranteed by the Talmud itself. As you will soon see, Judaism imposes no limits to one's interpretations, only to one's actions. This disclaimer also applies to our discussion of science in Part III, Texts of Science; the views are personal and do not necessarily reflect scholarly opinion. Indeed, all texts are personal, even scholarly texts. The reader, like the writer, has a uniquely self-organized (and self-organizing) mind [§1]; just as the writer, through writing, fashions a personal text, the reader, who wants to understand what he or she reads, must interpret (essential rewrite) the read-text in the light of his or her own mind. So reader, you too will have to make your own interpretations of this text. Our coming discussion of Talmudic texts will prepare us to discuss in more detail below *interpretation* [§68], *understanding* [§69] and *meaning* [§76].

We shall begin with Talmud texts that define the individual, the Person.

PERSON

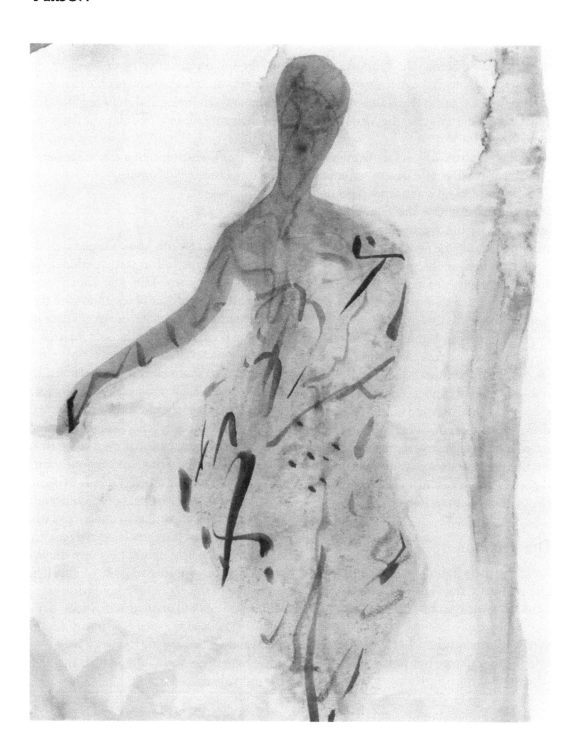

Subject One: Person

Here are some Talmud texts that explore the concept of individuality. Every human is the center of a universe and a special creation. But every individual lives within a society, a collective of individuals. Both the individual and the collective have rights and responsibilities.

§13 Rain and resurrection

Let us start with the title, here is a short text about the natural, the supernatural, and the person.

> Rabbi Abbahu said: Greater than the resurrection of the dead is the day of rain; for the resurrection is for the righteous only, while the rain is for both the righteous and the wicked. (Taanit, 7a)

These few words, there are only fifteen words in the Hebrew, appear with the cadence and rhyme of a poem. Like a haiku, the words of Abbahu embrace a worldview that is both human and down to earth. The rain—the natural regenerator of life in the dry Middle East—is superior to the resurrection of the dead—a supernatural act of God. And why? Because rain is not judgmental. In contrast to the resurrection, rain does not discriminate between individuals; the rain brings life to both the righteous and the wicked.

But, hold on. Is it not the aim of religion to separate the good from the bad by just reward and punishment? Is not God more glorious in the supernatural than in the natural? The answer, says Abbahu, is *no* to both questions.

Abbahu lived in Caesarea about 300 CE, a center of Roman rule and Christianity in Israel.

His disdain for the resurrection of the elect must have annoyed his Christian neighbors, but it did not perturb his fellow Rabbis. The text that follows Abbahu's haiku comments dryly:

> Abbahu's statement, the sages point out, can be interpreted to be at variance with the statement of Rav Yosef, who explained that the rain must be equal to (but not greater than) the resurrection because the rain and the resurrection are mentioned side by side in the prayer book.

The sages don't seem to be worried about the gift of rain to the wicked.

Abbahu's Talmudic rain text does not refer openly to a more fundamental Torah text, but the Torah text is there all the same. Deuteronomy 11:14, which is recited by observant Jews every morning and every evening [§39], proclaims that a timely measure of rain is God's reward to the righteous; the rain is the resurrection of the earth. The prayer book's association of the rain with the resurrection of the dead was probably based on the text in Deuteronomy. But see how Abbahu's Talmudic interpretation raises the Torah text of Deuteronomy 11:14 to a new height. True, the rain, like the resurrection, is earned by the deeds of the righteous (the Torah text), but the rain is greater than the resurrection because, once earned, the rain benefits every man and every women (the Talmud text). The reward of the good individual is the individual's success in benefiting all individuals, even the wicked ones. Being good is one's responsibility to the collective. Besides, today's sinner could be tomorrow's saint.

Where is the functional level, the *Halakhah* code? We have already seen it presented in the Talmud by Rav Yosef: the rain is to be mentioned in the prayers alongside the resurrection.

§14 Abel's blood

Our first text centered on rain, the water of life. Now we shall pass to a liquid closer to the heart of the matter: blood. First let us look at the semantic text inherent in the Hebrew word for *blood, dam*. The root is D-M, and that connects blood to *adam (man*, in fact the first man), and to *adama, earth*, and to *adom, red*, and to *demama, silence*. The semantic relations of the roots of Hebrew provide an insight into cognitive linguistics, and a poem too.

The special status of the individual is made explicit in the way a court uses the imagery of blood to warn potential witnesses testifying in a case where capital punishment is at stake. The witnesses are burdened by the court (*Halakhah*) with responsibility for the life of the condemned using a Talmudic interpretation of a Torah verse. See how the three levels of text are woven together to generate a new view of the individual in the context of creation. The Mishnah in Sanhedrin, 37a and 37b relates the following:

> How do the judges caution the witnesses before they testify before the court in a murder trial? The court actually threatens the witnesses like this:
> Be sure you testify only to that which you know first hand; make no testimony based on circumstantial evidence or hearsay. Keep in mind that we shall cross-examine you on every detail of your testimony.
> Know that capital crimes are different in essence from crimes against property: In a crime involving property, a man who gives mistaken testimony need only pay back the money and be done with it. But in a capital crime, the blood of the accused and the blood of all of his potential descendants till the end of time hang on the words of the witness.

The court shoulders the witnesses with a burden of responsibility for the life of the human standing trial. As the text continues, the lesson is made by an interpretation of the word *bloods*, which appears in the first murder reported by the Torah.

> When Cain murdered his brother Abel, Scripture states that "Your brother's bloods are crying out" (Genesis, 4:10). Scripture does not say " your brother's blood" in the singular, Scripture says "your brother's bloods" in the plural; the use of the plural refers not only to your brother's blood, but also to the blood of his descendants for all time.

(Know that the Hebrew text of the Bible uses the plural *bloods*; don't be misled by the King James translation which uses the singular "thy brother's blood".)

The witnesses are warned, using a grammatical twist, that they are responsible for the yet unborn generations who will not come to life, if the defendant is judged guilty and executed. It is a fact of nature that people have children; it is a Talmudic interpretation of Torah that imposes personal responsibility onto nature.

The text of the Talmud makes a short diversion and then returns to the issue at hand, abruptly switching from Cain and Abel (and murder) to their father Adam (and creation), and from thence to every person of us.

> For this reason, Adam was created as a single individual, to teach us that whoever destroys a single human life destroys, as it were, a whole universe. And whoever sustains a single human life sustains, as it were, a whole universe.

A human life, even that of a suspected murderer, is equal to all of creation. Each of us is the center of the only universe any person can know—the universe centered in one's own awareness. The loss of a person is the loss of a world—anti-creation.

The text of the Mishnah continues with a chain of associations that are drawn from the singularity of Adam the individual.

> Adam was created as a singularity to establish peace among humankind, so that one may not say to his fellow "my father is greater than your father".

Adam is a metaphor for human equality, as well as the prototype for the irreplaceable individual.

> And thus to argue against the belief in many gods.

The singularity of Adam announces the singularity of God. The individual person signifies the Individuality of God, the person's creator.

> And thus to proclaim the greatness of the Blessed Is He: The emperor presses many coins from a single mould, and each coin is identical to the original. But the King of Kings Blessed Is He impresses each individual from the mould of Adam, and not one human is identical to any other.

The individuality and diversity of humankind is the glory of God. Any Caesar can make copies; only God makes an infinite diversity of worlds. And then comes the ultimate lesson of creation.

> Therefore, each and every one is obliged to say: for me was the world created.

Each human, irrespective of nation, race, rank, office, wealth, family, creed, or degree of righteousness is the center of a world and the reason for the creation of that world. This is the essence of the warning to the witnesses. You judge a person, and you judge a world.

(The spirit and the message of Sanhedrin, 37a,b is universal; all, Jews and non-Jews, are the children of the singular Adam. For this reason, I have chosen to translate a Talmudic manuscript that does not mention "Israel" in the text. You will find that the printed version of the Talmud states "*And whoever sustains the life of a single person OF ISRAEL*

sustains, as it were, a whole universe". The mention of Israel in the context of Adam is clearly a contradiction of the sense of the Talmud here. I imagine that some of the descendants of the Rabbis who wrote the Talmud found it difficult to accept the universality of the original text, and so the added word *Israel* made its way into the printed text. Of course, we don't know whether this word was added by the first printer or was already there in the manuscript he chose to print. Does the added word *Israel* detract from the printed text? No, I think the gloss, by its patent contradiction of the rest of the text, only highlights the universality of the message. The gloss also highlights how difficult it often is to make a consistent interpretation of an open text. We tend to read what we want to believe.)

After grandeur comes business; the court has to examine the witnesses and make a decision. But by now the witnesses want to go home; they have been overwhelmed by the responsibility. The responsibility, nevertheless, must be borne.

> But if the witness gets cold feet; "what do I need this trouble for?" Let the witness note Scripture: "One who has witnessed or knows or has seen and does not tell will bear the sin" (Leviticus, 5:1).
> And if the witness wants to shirk this responsibility, let the witness again note Scripture: "The loss of the wicked is a good" (Proverbs, 11:10).

Here is the essential conflict between the individual, a murderer in this case, and the well-being of society. The murderer is the center of a universe, but society is created by law and order. Law must rule for all. The punishment of the condemned murderer is the rule of law. Below, we shall see the tension between the individual and the rights of society reappear in another context: one in which the deviant is not a wicked murderer, but is in fact a saint, confirmed by God.

The Talmud, because it proclaims the irreplaceable uniqueness of the individual person, is sharply aware of the dilemma of conflicting interests and desires. How are we to adjudicate conflicts between competing individuals and between individuals and collectives? Abbahu has ruled out righteousness and wickedness as absolute bench marks of human value [§13]. The court here has ruled out social rank, family, race, and creed. How are we to adjudicate worlds? Keep the question in mind, the Talmud will return to it again and again. As we shall see, the solution of the Talmud to the dilemma of conflicting rights is not righteousness or rank or affiliation, but consensus.

The Talmud, in elaborating the Mishnah's warning, goes on to clarify the meaning of "circumstantial evidence", as opposed to eye witness, which alone is admissible. Inference is not acceptable; a witness can report to the court only what he has seen with his own eyes. Science too is concerned with the weight of concrete evidence.

> The Rabbis asked, what does inference mean? The court cautions the witnesses thusly: If you saw the accused pursuing the victim into a ruin (where no other person is likely to be lurking), and you ran after them, and you found the accused holding a sword dripping blood and the victim in the throes of death—know that you have witnessed nothing (Sanhedrin, 37b).

The Talmud, like science, seeks concrete evidence. Interpretation is grounded most firmly on concrete evidence. Indeed, the evidence required by the Talmud for capital punishment is so stringent that the death penalty was effectively abolished by the Rabbis; a single execution in 70 years was considered the sign of a "murderous court" (Makot, 7a). Even a confession by the accused is not admissible evidence of murder. The Talmud, pre-empting Freud, apparently knew about the psychological need of some innocent people to admit guilt; the Talmud certainly knew about forced confessions extracted by Roman oppressors. The Torah orders the death penalty for a significant number of crimes; the Talmud essentially interprets them all out of existence. The Torah states the principle of "an eye for an eye" (Exodus, 21:24); the Talmud interprets the Torah to mean "the monetary value of an eye for an eye" (Baba Kama, 83b).

Crimes involving goods and property are judged in Talmudic courts by procedures entirely banned in capital courts: circumstantial evidence, self-incrimination, and contradictions in claims, are accepted along with hard evidence. Property courts and capital courts are even composed differently: Three laypersons can serve as judges in a property court, if agreed upon by the contestants. Capital crimes, in contrast to money matters, require a court of 23 expert judges (Sanhedrin, 2a,b); a fickle jury of peers is inconceivable. Property is only a temporary convenience; money comes and money goes. Courts that deal with money matters are like mediation teams. Life, unlike money, is for keeps, and needs a different keeper.

In the Anglo-American tradition, witnesses need only be sworn to tell the truth in preparation for testimony; the court extracts objective truth through questions and cross-questions. Why does the Talmud see fit to prepare the witnesses with dire threats? Why doesn't the Talmud, like we today, merely have the witness swear to tell the truth?

It would seem that the Talmud is less cognitively naïve then is the modern court. The Talmud knows that the human brain is not a passive receiver of bare facts; every observation is an interpretation. Every witness is a judge, every testimony a decision. Scientists, like other witnesses, need to recognize that expectations and bias distort observation.

We now begin to see the connections. The text in Sanhedrin, 37a, began with the preparation of the witnesses for true testimony, rose to proclaim the unconditional value of individual life, and returned to a concrete analysis of the evidence. The Talmud characteristically develops principles of great weight by asking questions of pragmatic utility. The sublime emerges from the mundane, and returns to the mundane. *Halakhah*, as we said, is walking the world.

§15 Another's blood

The threats voiced above by the court [§14] in "cautioning" the witnesses stress the absolute value of human life and bare the intrinsic conflict between the sanctity of the individual, even of a murderer, and society's debt to law. In the case below, the death sentence is not a court's decision, but yours, and it is not society that is to be so saved, but you yourself. In Sanhedrin, 74a, the Talmud relates the following:

> A man appeared before the sage Rabba with the following dilemma. The village tyrant had commanded the man: kill this other man; if you don't, I will kill you.
> Rabba said to the man: be killed, but do not kill the other man; how do we know that your blood is more red than his? Perhaps his blood is more red than yours?

Note, Rabba's decision in this Talmudic text is not presented as an interpretation of some fundamental Biblical commandment. "Be killed, but do not kill the other", is credited by Rabba to reason alone. Rabba needs no Torah text to buttress the argument; there is simply no criterion by which you might prove your own life to be intrinsically more valuable than that of another person. Each individual, as we have been taught in Sanhedrin, 37a, is the center of a universe [§14]. It is self-evident to Rabba that the blood of each of us is created equally red. The blood metaphor is fitting; recall the D-M root shared by man (*Adam*) and blood (*dam.*).

§16 Life and water

Here is another variation on the equality of life.

> Two people are walking through a desert with one jug of water between them. If both drink, both will die before reaching home. But if one alone drinks, that one will make it to safety. Ben Patira taught: It is best for both to drink, so that neither will see the death of his fellow. But then Rabbi Akiva changed the (Halakhah) ruling. It is written in the Torah (Leviticus, 25: 36) "so that your brother will live with you"; the one who possesses the water takes precedence. (Babba Metzia, 62a)

Since there is no reason to prefer one life over another, Ben Patira sentences both wanderers to an equal death; it is unfitting for a human to witness the death of a fellow human. Rabbi Akiva, however, wants a survivor; one life is more valuable than are two deaths. Since one life cannot take precedence over another, the owner of the water need not sacrifice his or her life. By the same reasoning, the one who does not possess the water has no claim on the life of the one who does possess the water. The one who possesses the water, let that one drink the water.

Akiva invokes a Torah text in Leviticus to support his rejection of Ben Patira's choice for death. But is the verse in Leviticus really relevant to Akiva's argument? What is the context of the Leviticus statement? Well, that Torah text contains nothing overtly related to the one jug of water for two wanderers. The verse in Leviticus, "so that your brother may live with you", is a justification for banning usury; support the needy with a loan, interest free, so that the needy can get back on his feet and live with you. But see, the common feature in both texts is life; the poor man needs to buy food to live, the man in the desert needs water to live. Akiva interprets the Torah text as a command for life; dividing the water means death. Again, who gets the water? The one who has the water.

Akiva's very choice of Torah text is a lesson in itself. One could find a number of suitable statements in the Torah which Akiva might have used to justify changing Ben Patira's *Halakhah* ruling; why use a commandment against usury to derive a person's right to use his property to save his own life? Akiva's choice of Torah text actually is an argument against any claim that the rich, the haves, enjoy a preference over the poor, the have-nots. The right use of property is to sustain life. The argument for life is based on the explicit commandment to use your property to help your brother get back on his feet, without extracting any "interest" for yourself (usury is forbidden; the loan is obligatory). Indeed, the case of the two wanderers and the jug of water is cited in the context of a Talmudic discussion of usury. The *montage* is meaningful; your right to save your own life with your own water, says Akiva, is only a special case of the generic obligation to sustain human life, which the Torah specifically puts in the context of sustaining the *other*. You are not superior to your poor brother, but, as Akiva now teaches, you are not inferior to him either; you may drink your own water. Property, we may infer, is only a means to life, not an end in itself.

Note that Akiva's association of the Torah text on usury with the story of the thirsty wanderers in the desert is a two-way connection. Just as the right use of water is equivalent to a loan, a loan given to a needy human may save the person's life. We are all wanderers in a worldly desert, and, to live, we must share our advantages with our fellows.

Contrast the Talmud with a similar dilemma discussed in Roman law: Two humans have survived a shipwreck in deep water (the stormy Mediterranean is the Roman desert). The two have come upon one plank of floating wood, and it is large enough to keep only one man afloat; if both try to hold on, the plank will go under and both will drown. Which of the two takes precedence? Roman law assigns the plank to the one whose life is more valuable to society, to the one of higher social rank.

The contrast is clear; the Romans distinguish individual life by individual power and social utility. The Romans, like the Rabbis, use logic to judge the value of the individual. But the two logics are not the same. The Romans measure the person's contribution to the harmony of society. Look at the person, they say, and his worth is obvious; worth is public service; worth is property. Not so the Talmud. The Talmud sees no logical standard for gauging individual life; blood is blood, equally red. Property is not a test of worth. Western democracy seems to have preferred the logic of the Talmud to the logic of the Romans.

§17 Individual name

The Talmud, in its characteristic manner, asks abstract questions through concrete examples. The Talmud, as we read above, sees the glory of God expressed in the individual diversity of people [§14]. What is an individual? Or, to put the question in functional terms, how do we recognize an individual when we meet one? Shortly after the description of the court's threat to the witnesses, Sanhedrin 38a analyzes individuality and quotes the following:

> Rabbi Meir says, one person differs from another in three ways: in voice, in appearance, and in cognition.

Rabbi Meir did not say that individuals are unique because each has a unique soul, or bears a unique genome, or owns private property; Rabbi Meir listed the ways we humans can witness uniqueness, by sensory perception and by experience. What we hear, even a brief hello on the telephone, can often suffice to identify a familiar person, and what we see can certainly suffice. Of course, we may meet genetically identical twins, or forget a face, or one person may impersonate another, but the twin, the impersonator and the original will always differ on the inside, cognitively. We know that from looking into ourselves. We know that even genetically identical twins possess different minds [§1].

We institutionalize individuality by assigning names, a name for each person. The cardinal importance of names in science is notorious; authorship and list of publications are vehicles of advancement. The order of names on a scientific paper can be a source of bitterness, scandal, feud, and shame, as well as of brotherhood. Scientists scan the citation index to see how often they have been quoted by name. Although science is a collective effort, we award prizes to individual scientists; we remember their names. Why? Because the progress of science emerges from individual excellence and from individual daring. A conceptual breakthrough is always a contradiction of the accepted dogma; the individual is the vehicle of progress.

The individual is recognized by name in the Talmud too; the names of the authors of each statement are carefully recorded. Both primary and secondary sources are given in detail. "Rabbi X said" and "Rabbi X reported the following statement from Rabbi Y". The Talmud, before editing, was preserved in memory, and memory can fail: "Rabbi X and Rabbi Y disagreed about a certain issue, but we don't recall which side of the dispute each took". "Rabbi X said in the name of Rabbi Y, and some say it was Rabbi A in the name of Rabbi B". The Talmud, like the cinema, is meticulous about credits.

Each person is the center of creation; whoever interprets a text in a new way creates an individual world. The texts of mythology name kings and queens, knights, priests and gods. The texts of science, like the texts of the Talmud, name every individual who has made a contribution. The Talmud is obsessed with names. Avot; 6,6 states,

> Whoever cites a text in the name of its author redeems the world.

The individual lives in his or her text. As we shall see below, even God has a private name.

§18 Weaving texts

By now, you should begin to get the feel of the Talmudic method. The Torah, the fundamental text is immutable and ever present; the Talmud exploits fragments of this text to construct its own meta-text: We saw how the Talmud used selected fragments from the Genesis story of Cain and Abel and Adam [§14]; the un-cited but implied rain text of Deuteronomy [§13]; the usury text of Leviticus [§16]; the citations from Proverbs and again from Leviticus [§14]. The Rabbis mine raw material from the Torah to construct Talmudic principles and *Halakhah* applications, all to provide a program for human life. Teach (Torah), learn (Talmud) and go (*Halakhah*). The Talmud, like science, has evolved organically over time [§6]. Its text is a collective fabric woven by individuals. In the Talmud, as in science:

> Jealous competition among the learned generates knowledge. (Baba Batra, 21a)

Just as the action needs personae, the drama needs a setting.

PLACE

Subject Two: Place

Place is a name of God. Place is a part of space defined by events. Place is the aspect of nature accessible to humans. We define the meaning of "holy" and discuss Talmudic texts that explore relationships between humans, nature, the supernatural, the devil and the creative urge.

§19 Place names

Place is a simple enough word in English, but not in Hebrew. The Hebrew word for *place* is *maqom*; the root is Q-W-M.

(Parenthetically, you might want to know why our Latin alphabet contains both a K and a Q, although they sound the same to us; Hebrew and Phoenician, from which the Greeks brought the alphabet into Europe [§9], pronounced the two letters differently, but only to their ears.)

The root Q-W-M serves also to create the words *qum* (arise, stand), *haqama* (establish), *tequma* (renaissance, revival), *qomem* (resist, stand firm), *qiyum* (existence), *yequm* (the universe). The Hebrew *place* is not a passive segment of space; the Hebrew *maqom* is existence carved out of space; it is the creation of a piece of space by an event. *Place* is the segment of nature accessible to humans; *place* is where we live.

There are a number of names for God that appear in the Torah, and we shall discuss them below in Section Three: Time. The Talmud adds several new terms for God, and prominent among these designations is *Ha-Maqom*, The Place.

Rav Huna said in the name of Rav Ami: why is the Holy One Blessed Is He called Place? Because He is the place of the world, but the world is not His place." (Bereshit Rabba, 68)

Rav Huna cites Rav Ami's interpretation of "Place": nature exists in God, but God exists outside of nature.

Rav Hanina son of Isi said: There are times when the universe and all its host is too scant to contain the glory of God, and there are times when He speaks to a person from between the hairs of his head. (Bereshit Rabba, 4)

The magnitude of a place is relative to events that occupy it, just as we have been taught by Einstein. Place is where each of us lives.

§20 Holy discrimination

The concept of Place, *maqom*, provides an opportunity to define the Hebrew concept of *qadosh*, usually translated as *holy*. The English word *holy* is related to the same root as the word *whole*. English *holy* is completeness, undifferentiated perfection, being one with reality, wholeness. Holy is the generality. Hebrew *qadosh*, in contrast, is being distinct, designated, different, dedicated, separate, special. *Qadosh* is the particular. The act of marriage is *qiddushin*, the designation of an exclusive relationship between a man and a woman. In Hebrew, we make something holy by relating to it, by distinguishing it positively from all the rest, by individualizing it with special obligations and rights. A holy place is a place set apart; a holy person is a person set apart; a holy day is time set apart [§43]. *Qadosh* (holy), like *maqom* (place), is defined by an exclusive relationship, at least in Hebrew. *Qadosh* demands individual separation, not wholeness. *Qadosh* is not The One; *qadosh* is The One among The Many.

§21 Akhnai's oven

Here is a text that weaves together God, nature, society, and the individual around a fanciful issue of ritual purity. The Talmud acknowledges that there are regulations to which any rational human society must adhere, like a ban on violence and murder, while other regulations are peculiar to the Torah (Yoma, 67b). Ritual purity, like the ban on the meat of pigs, belongs to the latter class. One would imagine that interpreting the irrational class would be more the prerogative of God and less the prerogative of the community. Yet, our case will demonstrate the opposite: the Rabbis will tell God not to interfere in the community's interpretation of the pure and the impure.

The particular issue under discussion has not been practiced in Judaism for 2000 years, and is a complicated subject even for experts. We only need know that someone named Akhnai has invented a type of oven that technically would be resistant to ritual impurity. The oven is constructed of broken pieces; the oven, if allowed, would evade impurity because a utensil may become ritually impure (by coming into contact with a dead body, for example) only if the utensil is intact. An artifact must be perfect to be susceptible to impurity. The capacity to become impure, then, is a sign of ultimate worth. What is not well done cannot be considered either pure or impure.

The contestants are God, Rabbi Eliezer, a respected conservative who often takes a minority position, and Rabbi Yehoshua, a liberal spokesman for the consensus. The point in question is whether functionality can determine

perfection. Rabbi Eliezer, the idealist, rules that Akhnai's fragmented oven is not worthy of impurity, while Rabbi Yehoshua, the pragmatist, and the majority claim otherwise; if the oven is sufficiently intact to cook with, it must be intact enough to become impure, broken pieces notwithstanding. Baba Metzia, 59b goes on to say:

> On that day, Rabbi Eliezer provided all the reasons in the world supporting his point of view, but they were not accepted by the Yeshiva.
> Eliezer said to the assembly: if the Halakhah regarding Akhnai's oven is as I say it is, then let that carob tree show me right. The carob tree promptly moved 100 cubits away. Some sources report that the tree actually moved 400 cubits. (A cubit is about half a meter).
> The Rabbis responded: a carob tree is no evidence.
> Eliezer persisted: if I am right, the aqueduct will prove it. The water in the aqueduct began to flow up-hill.
> The Rabbis responded: an aqueduct is no evidence.
> Eliezer persisted: if I am right, the walls of the Yeshiva will prove it. The walls of the Yeshiva began to cave in. Rabbi Yehoshua rebuked the walls: When sages are disputing the Halakhah, what business is it of yours? The walls, out of respect for Rabbi Yehoshua, could not fall, but, out of respect for Rabbi Eliezer, neither could they rise again. So the walls still hang suspended in midair.
> Eliezer persisted: If I am right about the Halakhah, let Heaven prove it. A Voice from Heaven said: Why are you contending with Eliezer? The Halakhah is as he says in all matters.

(God would seem to favor the idealist position here.)

> Yehoshua then rose to his feet and said: The Torah states "It is not in heaven" (Deuteronomy, 30: 12). What does the Torah mean, "not in heaven"? Rabbi Yirmiya has provided the answer: since the Torah has already been given to us, we no longer need take note of Voices from Heaven. You God have already written in Your Torah "follow the majority" (Exodus, 23: 2).

Here the majesty of the Hebrew is replaced by colloquial Aramaic, and the scene abruptly shifts

Consensus

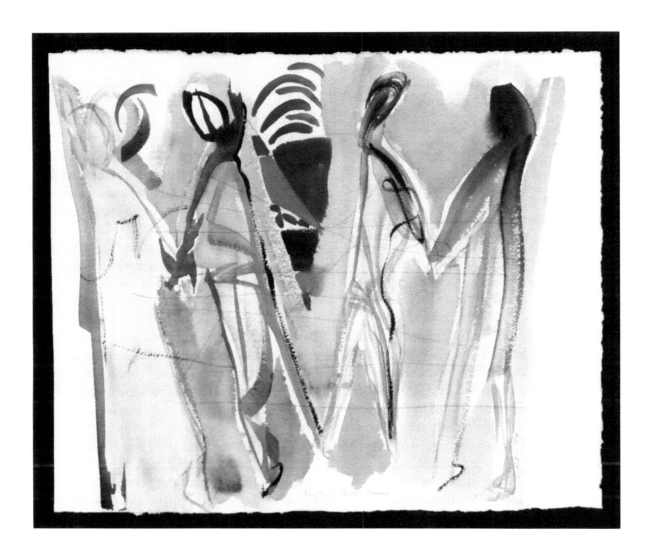

from the Yeshiva to the market place. Rabbi Natan chances to recognize Elijah the prophet in the crowd.

(Elijah is an emissary who often reports to earth developments in Heaven. You may recall that Elijah at the end of his career did not die, but was carried off in a fiery chariot as reported in the book of Kings II, 2: 11. Since he did not die, Elijah can return to earth, presumably using his chariot. In the Christian tradition, John the Baptist is a manifestation of Elijah.)

> Rabbi Natan asks Elijah: What did the Holy One Blessed is He do at that moment?
> Elijah said: God chuckled and said:

(The text shifts back to Hebrew; God speaks Hebrew, not Aramaic).

> My children have vanquished Me, my children have vanquished Me.

The messages here are clear. Truth is not to be had by reading the mind of God. The miraculous is not relevant to human affairs; the interpretation of the Torah is the provenance of knowledgeable humans who know the rules of interpretation. The meaning of the text is independent of the intentions of its Author; the true interpretation of a text is created by consensus. A lone visionary like Eliezer may see a truth, but such truth carries no weight in society, unless the lone visionary can convince the majority.

The text of Aknai's oven is often cited as an example of enlightened rationalism; see *Old Wine, New Flasks: Reflections on Science and Jewish Tradition*, by Roald Hoffmann and Shira Leibowitz Schmidt, W. H. Freeman, New York, 1997. Indeed, one could replace the word *Torah* by the word *nature* and withdraw the Author, and my summary of the text in Baba Metzia 59b is an apt description of the community of science. Here is

typical Talmudic chutzpa; Yehoshua uses God's own Torah to reject the legality of God's intervention in *Halakhah* affairs; the Torah itself has already endorsed the rule of consensus.

You may be interested to learn that the endorsement of the majority opinion, "follow the majority", in Yehoshua's rebuttal to the Voice contains a concealed sub-text; go to Exodus 23: 2 and see what the quoted verse really says. You will find that the words excised by the Talmud from Exodus 23:2 are only a fragment of the Torah verse. The complete verse actually says that one should *not* follow a majority bent on evil or conflict. The Talmud takes the words "follow the majority" out of the context of the Torah text and converts the original *negative* teaching into a *positive* teaching.

By citing words from the Torah out-of-context, the Talmud in effect is proposing an interpretation of the Torah and a commentary on majority rule. By saying "do not follow a majority bent on evil", the Torah text is interpreted by the Talmud to mean: do not follow an evil majority, but do follow a righteous majority. This interpretation is the basis of Yehoshua's argument against the Voice; a Yeshiva majority can justly reject Eliezer's opinion.

In citing Exodus 23: 2, the Talmud also sends a clear message that the power of the majority is not absolute; right and justice take precedence over the tyranny of an evil majority. The rub is that Eliezer is endorsed by the Voice from Heaven. So why not reject the Yeshiva majority and accept Eliezer as the just minority? The Talmud raises a complex question of conflicting rights, and leaves us to our own interpretation. I would say that the majority in the Yeshiva, like the lone Eliezer, are just and not evil. In a case where neither side is evil, go with the majority, the consensus. The individual, however, is commanded not to follow a majority bent on evil. More than one tyrant has been democratically elected. The individual is responsible for determining the difference between good and evil; the individual cannot delegate his or her moral responsibility to the majority.

We can conclude another important lesson from Yehoshua's argument, a lesson from what he does not say. Yehoshua and the Talmud do not get themselves embroiled in the question of whether miracles or Voices from Heaven do or do not exist in this world. The approach of the Talmud is existential and instrumental; what is the significance of such events in human affairs, that is the question. This world is a complex place, and people do hear voices from heaven and do attest to miracles. A Voice from Heaven in Hebrew is called *bat kol*—a *daughter of a voice*, which could be an echo, an intimation, an inner voice. The Talmud does not question the phenomena; the Talmud only concludes that such phenomena have no place in the normative standards of human society. Indeed, one might view the rejection of Eliezer's miracles as a veiled Talmudic critique of the emphasis placed by Christian doctrine on miracles.

Most of those who cite the disputation about Akhnai's oven stop here, basking in rational satisfaction. But the continuation of the story is dark; what befalls the individual protagonists after God's chuckle shows us the complexity and paradox of human life. The Talmud continues in Hebrew:

> On that day all the purifications done by Eliezer were annulled and the artifacts burned in fire. And the Rabbis voted to put a ban on Eliezer, to expel him from their company.
> But who could be entrusted to tell Eliezer of his expulsion?
> Rabbi Akiva, who was a student of Eliezer, said to the assembly: I will go tell him, lest an unworthy messenger do so and thereby destroy the world.

The issue has now shifted from society to the individual; Eliezer the man has been declared wrong by his colleagues, though shown by God to be right. God may have conceded the issue to the assembly with a chuckle, but Eliezer will stand firm to the edge of doom. Akiva, whom we saw to be a champion of life in the case of the thirsty wanderers [§16], has volunteered to soften the blow to the expelled dissenter.

> What did Akiva do? He clothed himself in black and enwrapped himself in a black Talit (a prayer shawl or toga), and sat himself before Eliezer at a distance of four cubits.
> (A banned person is not to be approached within a distance of four cubits.)
> Eliezer said: Akiva, what is unusual today? (Why black? Why four cubits?)
> Akiva said to him: My teacher, I think that the colleagues have separated themselves from you.

Akiva addresses Eliezer as his teacher, despite the ban, and frames the expulsion of Eliezer from the community as if the community has expelled itself from Eliezer. But to no avail.

> Eliezer too rent his cloths, removed his shoes and fell to the ground, in mourning.
> Tears fell from his eyes, and the world was shaken: a third of the olive crop was lost, a third of the wheat harvest was lost, a third of the oat harvest was lost. And some say, even the dough in the hands of the woman preparing bread for her family went sour.
> A great anger struck on that day; any place upon which fell the glance of Rabbi Eliezer was consumed in fire.

The joke is over.

Two additional persons now enter the scene. The first is Rabban Gamliel. (*Rabban*—our master—is a grander title than is mere *Rabbi*—my master.) Rabban Gamliel, a descendent of King David, is the president of the assembly, and so is responsible for the expulsion of Eliezer. The second person is Ima-Shalom (Mother-of-Peace). Ima-Shalom is the wife of Eliezer and the sister of Gamliel; Eliezer and Gamliel are brothers-in-law. The issue becomes a family triangle, and it is the woman, as usual, who is charged with keeping peace. The text continues and the scene shifts again:

> Rabban Gamliel was that day on a ship on the high seas, and there appeared a great wave that threatened to crush the ship. Gamliel said to himself: Such a wave could only come because of Rabbi Eliezer the son of Hyrkanos.
>
> Gamliel rose to his feet and said: Master of the Universe, You surely know that I acted, not for my own honor or for that of my family, but only for Your honor, so that there be no dissension in Israel.
> The sea ceased its raging.

The Hebrew phrase "the sea ceased its raging" is almost identical to a passage in the Book of Jonah describing the return of calm after Jonah is cast into the sea by his shipmates (Jonah 1: 15); the description of the sea in the Talmud is a veiled reference to Jonah in the Bible. Jonah, like Rabbi Eliezer, was an intransigent dissenter [§61]. The Talmud, in its choice of words, implies that Eliezer, like the prophet Jonah before him, had to be cast out for the well-being of the collective.

The scene now shifts to Ima-Shalom and her house. The language shifts from Biblical Hebrew to household Aramaic.

> Ima-Shalom was the wife of Rabbi Eliezer and the sister of Rabban Gamliel. From that day on, she prevented her husband Eliezer from falling to the ground to say the prayer of anguish.

The Jewish prayer book contains a prayer in which a person expresses sorrow and anguish. Typically the person so praying falls to the ground or, more commonly now, buries his head in his arms. The prayer of anguish is not said on Sabbath days or holidays, when rejoicing is the tone. Ima-Shalom feared the consequences of her husband venting his anguish, and so prevented him from falling to the ground in the prayer of anguish.

But one day, Ima-Shalom missed her duty because she mistakenly thought that the new-moon festival was a two-day holiday, while it was only a one-day holiday.

The new moon is marked by special festive prayers, and the prayer of anguish is forbidden. Some months the new moon is celebrated for two days and some months for one day only. Ima-Shalom mistook a one-day new-moon festival for a two-day festival, and neglected to supervise her husband's prayers on what she thought was the second day of the new moon.

> Some say that it was not the new moon, but that Ima-Shalom left her station to give bread to a poor beggar at her door.
> She returned to find Eliezer falling to the ground to express his anguish.
> Arise, she said. You have killed my brother.
> Too late. The blast of the ram's horn was already heard announcing the death of Rabban Gamliel.
> Eliezer turned to his wife: How did you know that this was going to happen?
> She said (switching from Aramaic to Hebrew): I learned in my grandfather's house that the gates of Heaven may be closed to prayer, but never to anguish.

What are we to make of this tragedy? Is there a hero? Is there a villain? The Talmudic script dramatizes a primal conflict of the human heart—the death of a brother caused by a brother-in-law's anguish and a sister's absence of mind. Who is guilty? How do Ima-Shalom and Eliezer now proceed through their remaining life together? Disaster visits a family, though each of the three is trying with heart and soul to do the right thing.

Note that the word Akhnai is related to the Hebrew word *akhan*, which means *serpent*. The term *Akhnai's oven* means *serpent's oven*. Thus Akhnai's oven could, as we said, refer to the man named serpent who invented the oven. The word Akhnai could also refer to the serpentine form of the fragments from which the oven was made. But in calling the oven the *serpent's oven*, might not the Talmud be invoking some association with the earlier triangle in Eden involving a man, a woman and a discussion with God (Genesis 3:1-21)?

Two conclusions may be drawn from the story of serpent's oven. The anguish of the human heart has an impact on the conduct of the world, and the individual, even a saint like Rabbi Eliezer, must be aware of the cost of self-righteous intransigence to himself and to his world. The first part of our story, the part taking place in the public forum of the Yeshiva, is all reason, smile and light; the second part is the dark complexion of the human heart. And the world attends to both.

Indeed, the text is dialectical; the second part is the antithesis of the first part in language (Aramaic versus Hebrew), in humor (dark versus light), in staging (individuals alone versus public assembly), in rationale (mystery versus demystification), in discourse (anguish versus argument), in society (forum versus family), in gender (woman versus men), and in outcome (death versus victory). Akhnai's oven is a dynamic composition in point and counterpoint. Cinematic exposition at its best. The Talmud, true to life itself, glories in contradiction and thrives on uncertainty. Clarification is the synthesis that emerges through the give and take of contending truths. "Both positions are the words of the living God" (Eruvin, 13b [§67]).

Rabbi Eliezer should have followed the example of his protagonist Rabbi Yehoshua. Yom Kippur, the Day of Atonement, is one of the holiest of days [§61], and according to Yehoshua's reckoning the community was planning to inaugurate Yom Kippur on the wrong day. The assembly refused to accept Yehoshua's calculations, and Rabban Gamliel commanded Yehoshua to desecrate his own calculated day of Yom Kippur by appearing before the assembly with his staff and money belt. Yehoshua accepted the consensus ruling, and that was that (Rosh Hashanah, 24b). No miraculous invocations; no voices from heaven; no death and disaster. But Eliezer will not bend.

§22 Rabbi Eliezer, Jacob of Kfar Sekhaniah and Jesus of Nazareth

Rabbi Eliezer embodies a contraposition to normative Talmudic thought. Eliezer is a miracle worker who clings to a private interpretation of the *Halakhah* regarding Akhnai's oven [§21], and other matters too. And so Eliezer becomes a Talmudic outcast. His departure from the consensus was clear even to the Romans. Eliezer lived through the destruction of the Second Temple (70 CE) and into the beginning of the second century—a time when Christianity was outlawed by the Romans as a subversive superstition. Eliezer, the outcast, was arrested by the Romans on suspicion of being a Christian. Avodah Zara, 16b, tells the story.

> When Rabbi Eliezer was charged with being a Christian, they took him to the place of execution.
> The Roman Proconsul said to him; How can a respected elder like you be caught up with such foolishness? Eliezer answered him; I place my trust in The Judge. The Proconsul assumed that Eliezer was referring to him. But Eliezer was referring to his Heavenly Father. Said the Proconsul: Since you place your trust in me, I shall set you free. Dismissed.

Eliezer, however, was distressed by the whole affair: his arrest and his freedom by the whim of a vain Proconsul.

> When Eliezer reached home, his disciples gathered to comfort him. But he would not be comforted. Rabbi Akiva said to him; My master, allow me to repeat one of the lessons you taught me.

Said Eliezer; Speak.

Said Akiva; My master, perhaps you once had contact with a Christian heresy and found pleasure in it? And for that you were arrested as a Christian.

Rabbi Akiva again came to comfort his teacher, and Eliezer recalled that he had once found pleasure in a Christian interpretation of a *Halakhah* that he had heard in the name of Jesus of Nazareth. (The name *Jesus of Nazareth* has been censored from certain editions of the Talmud; my citation is from the Steinsaltz Edition of the Talmud, cited above). The story continues:

Said Eliezer; Akiva, you have reminded me. Once I was strolling in the upper market of Sephoris (a town in Galilee), and I happened upon one of the disciples of Jesus of Nazareth, Jacob of Kfar Sekhaniah was his name.

Jacob said to me; It says in your Torah: You shall not bring the wages of a prostitute.... into the House of God (Deuteronomy 23:19).

What about using such wages to build a toilet for the High Priest?

The question sounds frivolous, if not disrespectful; but the question is serious. Sacred prostitution was an important ritual in the fertility rites of the natural religions that preceded Judaism (see *The Golden Bough. A Study in Magic and Religion*, by J. G. Frazer, 1 Volume, Abridged Edition, Collier Books, New York, 1963 edition of the original of 1922). Pagan temples throughout the Mediterranean world and Near East were maintained by the wages of sacred prostitutes who represented the great mother, the goddess of fertility in her various incarnations. The ban on fertility rites was a notable innovation of the Torah and a milestone in the transition from the sacredness of nature to the sacredness of history (more below [§37]). The High Priest is obliged to live in the precincts of the Temple in Jerusalem for a week before his service on the holiday of Yom Kippur (Yoma, 2a); he surely needs a toilet. Defecation, like the rite of love, is an important natural function. The love-bed is an enactment of creation and the toilet receives the residue of destruction; might not the two acts—construction and destruction—be joined naturally? The cycles of nature, reflected in the cycles of mythology, are cycles of procreation and doom [§36]. Jacob's question, which connects the two acts, is logical in this light. Eliezer is struck speechless.

I said nothing to him.

Said Jacob to me; Thus was I taught by Jesus of Nazareth (citing Micah 1:7): Wages originating in filth should return to filth.

And I found pleasure in this interpretation. For this was I charged with Christianity.

We may infer from Jacob's report of Jesus' Talmud-like question and answer that Jesus taught his disciples that the two acts—love and excretion—are indeed joined; but, by pure ritual, we can rise above nature's rubbish.

Eliezer's familiarity with Jacob the Christian and Eliezer's pleasure in Jesus' *Halakhah* were understood by Eliezer to have been the cause for his punishment by capricious Roman power. The proconsul was capricious, but God is never capricious. The incident in the Upper Market, according to Eliezer, justifies the act of the Heavenly Judge.

Eliezer, despite his ban from the Yeshiva, remained in the fold, and, at his death (from illness), he and his teachings were reinstated into the normative mainstream by his Talmudic colleagues (the story is told in Sanhedrin, 68a). Eliezer was not put to death as a Christian by the Romans, but Akiva, his disciple, was killed some years later by the Romans for teaching Talmudic Judaism (Brakhot, 61b).

This text connecting Rabbi Eliezer with a point of *Halakhah* attributed to Jesus is intriguing. If the Talmud here is historically accurate, we may conclude that Jesus and his disciples used a form of Talmudic discourse; they spoke the hermeneutic language of Eliezer and his colleagues. The text could be interpreted to imply that the historical Jesus was perceived to be a Rabbi, but one whose *Halakhah* teaching, like that of Eliezer, was outside the Talmudic consensus. For good reason did the pagan Roman authority suspect Eliezer of being a Christian; Eliezer, like Jesus, was a dissident Rabbi. Eliezer's indulgence in miracles certainly did not contradict this Roman misconception (see above [21] and Sanhedrin, 68a).

A description of life in the world in which Christianity developed can be found in *The New Testament Environment*, by Eduard Lohse, Abingdon Press, 1976. To learn about the divergence of Christianity from Judaism after Jesus, you may want to read *Constantine's Sword: the Church and the Jews: a History*, by James Carroll, Houghton Mifflin, Boston, 2001.

(The triangle of confused identity—Jewish-Christian-Roman—persisted and is illustrated by the interpretation provided by Rashi to the story of Eliezer's trial—in Avodah Zara, 16b. Rashi, the authoritative Jewish commentator [§6], lived about a thousand years after Eliezer, at the time of the Crusades in what is now Mainz. By then, Rome was not pagan but Christian; Christianity was no longer a Jewish sect, but a separate religion; and Roman persecution, preached by the Church, was directed, not against Christians, but against Muslims and Jews. Rashi simply cannot believe the text of Avodah Zara, 16b, as it is written; Rashi assumes that Eliezer was tried, not by pagan Romans on suspicion of being a Christian, but by the Christians, to force his conversion from Judaism. Eliezer's judge in the Talmud story is termed a *hegmon*, originally a Greek word for *leader* or *ruler*. The Talmud used the term to describe a Roman Proconsul. But by the time of Rashi, *hegmon* had come to mean a Bishop of the Catholic Church. Each generation and its misconceived identities.)

Eliezer and Jacob (citing Jesus) would seem to agree that rites of fertility, however essential to nature, are, like the toilet, foreign to the Temple ritual. However, the toilet and the bedroom belong to creation and so are still to be studied as Torah—God's text.

§23 In the toilet and under the bed

Rabbi Akiva said: One time I followed Rabbi Yehoshua into the toilet and I learned from him three things: that one is to face North and South rather than East and West; that one should wipe sitting and not standing; and that one should wipe with the left hand rather than with the right hand.
The Son of Azai said: How dare you show your master such impudence!
Said Akiva to the Son of Azai: It is Torah, and learn I must (Brakhot, 62a).

The Talmud goes on to tell us that the Son of Azai undertook the same study and followed, in his turn, Akiva his teacher into the toilet to confirm the three lessons. When his deed was challenged, in turn, by Rabbi Yehuda, the Son of Azai gave the same answer he heard from Akiva: It is Torah, and learn I must.

The text then switches from Hebrew to colloquial Aramaic, and the action from the toilet to the bedroom. The teacher here is Rav, a master of such renown that his name is simply Rav—*master*; there is no need to use his personal name.

Rav Kahana slipped under the bed of Rav, and heard him making love to his wife: speaking together, laughing, and doing it.
Kahana (under the bed) said to Rav: Father, you are as one who eats with a great hunger.
Rav said: Kahana! Is that you? Out! You desecrate nature.
Said Kahana: It is Torah, and learn I must.

CREATIVE URGE

2009   Venice   B. Rubin

The world, according to the Talmud, is not divisible into the mundane and the sublime, or separable into the worthy and the unworthy. Nature is Torah; nature is within God, *The Place* [§19]. And so all of nature is to be studied and interpreted. Every feature of the world is a signifier [§5]. The act of excretion and the act of love, destruction and creation, are Torah—acts, through learning, become teachings, and learn we must.

Note that the Son of Azai did not depend on Akiva's report of the lesson he claimed to have learned from Yehoshua; Son of Azai did his own study and followed his teacher Akiva into the toilet. The lessons of love learned by Kahana from Rav would seem to be communication, joy and passion; how may we interpret the three lessons from the toilet? Even excretion is a human ritual that deserves gesture and intent, even if foreign to the Temple. Close-up realism.

Science, like the Talmud, extracts the sublime from the mundane; fruit flies (*Drosophila*), worms (*C. elegans*), and bowel bacteria (*E. coli*), unlock the secrets of life.

§24 Constraints on power

All of nature is Torah; what about the *id*, the drive for power and sexual gratification? Is lust too Torah? Bereshit Raba, 9;9 puts it this way:

> The Torah states: And God saw all that He did, and behold it was very good (Genesis, 1: 31). Rav Nahman said in the name of his father Shmuel: "Good" refers to the good urge; "very good" refers to lust. But how can lust be very good? It's a contradiction! Consider this: without lust, a man would never build a house, marry a woman, produce children, or strive in the markets of life.

Lust, the *id*, is essential to life in this world. The Talmud in Yoma, 69b illustrates the point,

as usual, by telling a story. The text tells us that Israel was able, without penalty, to expel idolatry from the Service to God. This success led the Rabbis to dare to control the sexual urge, the other source of sin.

> They said, since the hour is propitious, let us pray for the end of this sin too. They prayed, and behold, lust for sexual gratification was placed in their power. But a prophet said: Beware: if you kill this urge, you destroy the world.
> For three days they deliberated the fate of their captive lust, but a fresh egg (to prepare a medication) could not be found throughout the Land of Israel. (Nature had come to a halt; creative energy was dissipated.)
> They said: what can we do? If we kill the lust for sex, we shall destroy the world!

Without the *id*, even the chickens won't lay eggs. In the end, the Rabbis found a way to release lust to its natural function, yet blunted so that the urge for incest would be slightly disabled.

There is an important point here. Learned people do have the right to seek the power to intervene in nature. Some natural features of the world, like idolatry in this story, can be changed with impunity; and some seemingly problematic aspects of the world, like the *id* in this story, are essential to life. Wisdom is discriminating ahead of time between what we may change and what we may not change in nature. Three days were enough for the Rabbis to see the essence of the matter. We seem to need more time to learn the difference.

§25 Creative urge

The Hebrew word for lust, the urge, the *id* is a lesson in itself. The word is *yetzer*; its root is Y-TZ-R. Words with the same root are *yotzer* (creator, artist, potter), *yetzira* (creation, work of art), *yetzur* (creature), *totzeret* (product), *yitzur* (production). Lust can generate a creative force, if we know how to use it—sublimation. The text in Sukkah, 52a makes clear the importance of sublimation:

> The greater the person, the greater is his yetzer (urge).

How to sublimate one's lust for life; that, says the Talmud, is the challenge, the opportunity and the art of human existence.

§26 The devil and the urge

The devil in the Jewish tradition is not the powerful contender with God depicted in Christianity. The Jewish devil is not God's opponent, but God's servant. The devil is the lawyer for the prosecution in God's court. The Biblical devil appears in the book of Job to challenge God to try the faith of His servant Job (Job 1:6-13). In the Talmud, the devil is the personification of the human urge, the chief prosecutor, and the test of human quality. Qiddushin, 81a/b tells four stories, short cuts, about four holy men who have some difficulty dealing with their instincts.

> A group of young women (who had been ransomed by the community) were brought to the city of Nehardeah (a noted center of Talmudic learning in Babylon) after their release from captivity.

Women in Babylon were all too often carried off, raped and held for ransom by their captors. One can imagine the excitement that such a group of young women might arouse upon their return to the community: women at the interface between lust and decorum, violation and family. The women were placed in the care of a most holy man, Amram the Hasid, to ensure their safe return to their families. (Hasid is a title that designates a person of great piety.) The drama

opens with a wordless shooting script followed by a short dialogue:

> The women were taken to an upper story of the house of Amram the Hasid, and the ladder was removed (so that none could climb up to the women; the ladder, a transparently sexual image, is publicly put aside).
> But it happened that one woman passed by the stairwell and the light of her beauty filled the room. Rav Amram seized the ladder, which normally required ten men to lift into place, and placed it by himself. And began to climb up. Half-way to the top, he regained his self-control, and began to shout: Fire in the house of Amram!
> The disciples came forth to help with the fire, (but only found Amram on the ladder). The disciples said to their teacher: You have shamed us.
> Said Amram to the disciples: Better to be shamed by the House of Amram in this world, than to be shamed by him in the next world.
> Amram then foreswore his lust to leave, and the urge exited his body as a pillar of fire.
> Said Amram to his lust: See, you are fire, and I am flesh. And yet I got the better of you.

Amram, energized by public shame, put out his own fire; one's flesh can empower one's spirit—sublimation works. The next episode tells of another noted holy man.

> Rabbi Meir used to laugh at the weakness of transgressors. One day, the devil appeared to him as a woman on the far bank of a river. There was no ferry, but Rabbi Meir grabbed a rope and began to cross. When he got half-way, the urge left him.
> The devil said to Rabbi Meir: Had it not been declared in Heaven, "be wary of Rabbi Meir and his teaching", your life would not have been worth two cents.

Who does Heaven warn to "be wary", the devil, the reader, or Rabbi Meir himself? In any case, the devil here serves as the spokesperson for Heaven. The devil makes public Meir's obligation as a teacher.

The Talmud goes on to tell a similar story involving Rabbi Akiva, who we already met above.

Rabbi Akiva used to laugh at the weakness of transgressors. One day the devil appeared to him as a woman at the top of a palm tree. Akiva grasped the palm and began to climb. When he got half-way up, the urge left him. The devil said: Had it not been declared in Heaven, "be wary of Rabbi Akiva and his teaching", your life would not have been worth two cents.

Strange, the stories of Meir and Akiva would seem to be the same story; even the wording is the same. Why does the Talmud bother to repeat itself? The minor differences, in the way of the Talmud, must bear major meaning. I can think of two lessons: First, the same folly befell two wise men, Meir and Akiva; such folly, therefore, is not unique to a Meir or Akiva, but is common to all, even to the wise reader. Secondly, Akiva, a more saintly figure than Meir, was even more foolish than was Meir: ravishingly beautiful women are customarily on the far side of the river (or at the window of a car or ferryboat going in the other direction), but how does the beautiful woman get to the top of a palm tree? Akiva should have smelled the devil at work.

Rabbi Meir and Rabbi Akiva were central figures in the creation of the Talmud and did not lack moral awareness or courage. We saw above, for example, how Akiva took upon himself the task of informing Rabbi Eliezer of the ban put on him [§21]; we saw how Akiva was a champion of life [§16], how he related to all of creation, even to the toilet, as Torah [§23]. Akiva was tortured to death by the Romans, and he declared his love of God as he died [§38]. Akiva, as we shall see, was the pride of God, and God sent Moses a thousand years into the future to visit Akiva's Yeshiva [§54]. Rabbi Meir

too was a legendary figure. Yet both men were blinded by a flash of beauty and almost fell captive to lust. They were certainly stimulated to perform great physical feats; crossing a flowing river and climbing a tall palm tree adds sexual imagery to their physical acts. The urge of Akiva and Meir for the distant and unapproachable figure of a woman is set in counterpoint to the passages in the Talmud that tell us that both men were married to women of high achievement who strongly influenced their husbands' professional lives.

Why did the Talmud choose to publish the shame of these moral paragons? Is public shame a punishment for shaming others, for vanity, for pride, for contempt of natural lust? I can imagine that the devil appeared to Meir and to Akiva as an unusually attractive woman. Yet Amram the Pious was moved by a glance at the beauty of a real woman placed in his charge. Perhaps shame is not only punishment, but therapy; Amram harnessed the force of his shame before his disciples to conquer his lust.

The fourth episode shows us the devil working, not in the guise of an alluring woman, but in the guise of poor man knocking at the door; temptation has many guises. Shame plays a curative role in this story too.

> Each day Pleimo used to curse the devil by saying: An arrow in the eye of the devil.

Yom Kippur, the Day of Atonement, is a day of holiness and prayer marked by a total fast from sunset to the next nightfall [§61]. In preparation for the fast and the atonement, each family gathers before sunset for a final solemn meal. The mood is one of sanctity.

> One day, Yom Kippur eve, the devil appeared as a poor man and knocked at the door. They brought him bread.

The devil said: On such a day as this when all the world are inside, am I to remain on the outside?

They let him in the house and gave him the bread.

The devil said: On such a day as this when all the world are sitting at the table, am I to sit by myself?

They seated him among them at the table.

The devil sat at the table and covered his skin with boils and pus, and behaved in an exceedingly disgusting manner.

Pleimo said to him: Behave yourself.

The devil said: Bring me a cup.

They brought him a cup.

The devil coughed and spit his phlegm into the cup.

They became angered at him. The devil fell as if dead.

An uproar was heard in the community: Pleimo has killed a man. Pleimo has killed a man.

Pleimo ran from the house and hid himself in the outhouse toilet. The devil went after Pleimo and fell before him.

When the devil saw Pleimo's distress, he revealed to Pleimo his true identity.

The devil said to Pleimo: Why do you curse me thus daily?

Pleimo said to the devil: But what am I to say?

The devil said: You should say, "May God be angry with the devil".

Here is an artful but most perplexing tale. The evil inclination, the devil, can manifest itself even as a good deed, feeding a poor man, and at the most solemn of occasions, Yom Kippur eve. Perhaps Pleimo and his family were being sanctimonious and vain in cursing the devil, as if they were immune to the evil urge. Pleimo wants to have a holy communion, and it is ruined by an uninvited and disgusting interloper. Perhaps Pleimo should have brought the poor man in to sit at his holy table without the man having to demand equality. Perhaps Pleimo should have offered to help the man with his skin and lung disease. A good deed is a temptation to pride. To silence the devil, a truly good deed needs a truly good heart.

Finally, what is the difference between Pleimo's daily curse that hurt the devil's feelings ("an arrow in the devil's eye"), and the statement that the devil instructs him to say instead ("may God be angry with the devil")? I can offer the following: The *id* is part of nature, an essential part of nature. The urge is not to be cursed or belittled, but its energy is to be harnessed and sublimated as a force for the good. If we suffer from the itch, let the Lord of the Universe consider an alternative; the One who gave us this burden is the One who can take it away.

Indeed, the four stories can be seen as consecutive variations on a theme that includes the contempt of a holy man for the moral weakness of his community; the man's precipitous fall from spiritual virtue and his exhibition of unusual physical energy; the public revelation of the holy man's fall and his public shame; and the corrective force of dialogue with the community, which the holy man distained before his fall. These elements are expressed in different ways in each of the stories, but all are there:

1. Contempt for weakness: Meir and Akiva openly laugh at the weakness of sinners; sanctimonious Pleimo scorns the power of the devil; and Amram the Pious was sure the women would be safe with him; he had the heavy ladder removed to prevent others from reaching the women.
2. Great energy can spring from lust or shame: Amram grabs the ladder with the strength of ten men; Meir swims the current; Akiva climbs the tree; Pleimo runs for the outhouse.
3. Public censure: Amram arouses his shocked disciples; Meir and Akiva are shamed by the devil (and by us who read about their fall); Pleimo is called

a murderer by the community and flees his holy table for the pollution of the outhouse.

4. Community pressure strengthens the individual: Amram's shame before his disciples empowers Amram to quench his fire; Meir and Akiva are obliged by their roles as teachers; Pleimo is driven to confront the devil by the community who charges him with murder.

5. The lesson: Amram harnesses the power of his lust and shame to subdue his lust and shame; Meir and Akiva have no answer to the devil's taunt (we can believe that they will no longer laugh at the weakness of simple sinners; perhaps they will see the woman as an individual, not as a persona of their own devil); Amram seeks the advice of the devil he used to taunt; Amram has learned to be attentive to his own urges.

The four-episode montage makes each story a commentary on the other three; the four stories together create a fifth "meta-story". Akiva and his colleagues were great spiritual leaders despite their surprisingly human weakness; perhaps a person acquainted with his or her own weaknesses might still aspire to surprising greatness. After all, each is a center of creation [§14].

Note, however, that each protagonist learns and progresses spiritually through interaction with a community. The Talmud ranks social interaction and social responsiveness far higher than lonely contemplation. Amram the teacher learns through dialogue with his students. Saintly Akiva heeds devilishly public censure.

Let us return to the human use of the urge to create; what is our place in nature?

§27 Human creator

> Rava said: If the righteous would so wish, they could create the world. The Torah says: Only your sins separate you from God (Isaiah 59: 2).
> Rava created a man and sent him to Rabbi Zera. Zera spoke to the man, but the man did not answer him. Zera said to the man: You have been made by a colleague. Return to your dust (Sanhedrin, 65b).

The Torah text in Isaiah 59:2 says that sin *distances* man from God; but Rava interprets the word to mean not *distance* in metaphoric *space*, but *separation* in metaphoric *essence*. If a human were sufficiently righteous, that human could, like God, create a world. Just to prove the point, Rava creates a man. But the man is imperfect; he is a *golem*, and so cannot speak. Zera senses that the strange silent man is an artifact, and returns him to primal clay. Why did Zera destroy the golem made by Rava? Was it only because the golem was speechless, or did Zera sense danger, or perhaps blasphemy, in Rava's mechanical man? Scientists of today do make unnatural golems (robots, computer networks, weapons of mass destruction); like Zera, we have to decide what to do with them.

Golem stories appear in later Jewish literature too; a golem was reported to have been made by Rabbi Judah Loew in the 16th Century to protect the Jewish community in Prague. (For more about Golems, silent and talkative, see *Golem: Jewish Magical and Mystical Traditions on the Artificial Anthropoid* by Moshe Idel, State University of New York Press, 1999.) The Frankenstein monster is the latest literary golem. Robotics experts in the near future will, no doubt, do in silicone as well as Rava did in clay. The point for now is that the Talmud allows humankind the ultimate intervention, the creation of life. The Talmud hedges that power with righteousness; access to creative power springs from right action. Our present problem is that modern technology cannot discriminate between right and wrong; righteousness is no longer a prerequisite for successful creation. Knowledge and power are now merely commodities of exchange. The challenge facing society is how to append righteousness to technology. The Talmud, as usual, preempts our discussion. Sanhedrin, 65b continues:

> Rav Hanina and Rav Oshiya spent a Friday afternoon busy with the Book of Creation (sefer yetzira), and created for themselves a three-year old calf. And they ate the calf for the Sabbath meal.

The Book of Creation is a mysterious book that describes the power of language to create [§9]. Language is the foundation of human culture and of human power over nature [§69]. Since the days of the Talmud, knowledgeable humans have fashioned golems and other creatures through the use of language, the Book of Creation. Whether or not Hanina and Oshiya really created their wonderful calf, the Talmud is telling us here that whatever we do manage to create, it is we who are going to have to eat it, even on the Sabbath.

§28 Yeshiva in Heaven

The following story is told in Baba Metzia, 86a:

> Rav Kahana said: Rav Hama, the son of Hasa's daughter, told me that Rabba the son of Nahmani died because of religious persecution. An informer told the Persian governor that a certain Jew was causing a loss of tax revenues by preventing the employment of 12,000 workers each year for a month in the Spring and a month in the Fall.

The Jewish community in Babylon organized a kind of bi-annual sabbatical in which the entire people would retire for a month from their customary economic activities and devote their time to studying Torah and Talmud with the Rabbis. This Talmudic Open University bred universal literacy and gave a taste of intellectual life to working people. The Rabbis knew that public education is the responsibility of every intellectual elite; public education, in fact, is a most efficient way to renew and strengthen society's intellectual resource. But the Talmudic University also aroused the wrath of the Persian treasury, who felt the loss of taxable income. Perhaps the authorities were wary of too much education. Rabba, the Rabbi who taught us above that all blood is equally red [§15], was a key organizer of the study sessions and the authorities went after him.

> The king's agent searched for Rabba in vain. Rabba kept one step ahead of the agent, going from Pumpadita to Akra, and from Akra to Agma, and from Agma to Shehin, and from Shehin to Tzrifa, and from Tzrifa to Ena Demaim, and from Ena Demaim back to Pumpadita. (Places in Babylon; is this a shooting script to help us visualize the moves: by donkey, camel, river boat, on foot?)
> In Pumpadita, the agent caught up with him.
> It happened that the king's agent chanced to lodge in the same inn where Rabba was hiding.
> Unknowingly, the staff at the inn caused the agent to become sick.
> (The sickness is mysterious and difficult to translate; apparently, it was some sort of magical deed that was done to the agent.)
> They became greatly alarmed when they discovered that they had put a spell on an agent of the king.
> Rabba took charge and effected the agent's cure.
> The agent concluded that his benefactor must be Rabba himself; and he apprehended Rabba. The agent, however, told Rabba that he would not deliver Rabba to the king, even on pain of death. But if tortured, he would turn Rabba in.
> Rabba understood the hint, and escaped to Agma. There he sat himself upon the stump of a date palm and proceeded to study Torah.

This is the background of the plot.

The scene abruptly shifts out of this world, and we get a glimpse of a Yeshiva session that takes place in Heaven. Obviously, the language shifts from Babylonian Aramaic to Hebrew (they speak Hebrew in Heaven). We see that the Rabbis in Heaven continue their Talmudic disputations as they did when they were alive on earth, except that God also takes part in the discussions: realism becomes surrealism.

The case under discussion in Heaven, like Akhnai's oven on earth, relates to the rules of impurity. Certain skin lesions are diagnosed as "leprosy" (see Leviticus, 14) and render a person impure, requiring the person to undergo a type of ritual bathing (a baptism). The Biblical "leprosy" is not what modern dermatologists would call leprosy; we simply do not know what the Biblical disease would be called today. Nevertheless, the disease has clear signs. The Talmud tells us:

> If a white patch of skin precedes the emergence of a white hair in the patch, the person is impure. But if the white hair precedes the white patch of skin, the person is pure.

The Rabbis in the Talmud are required to know their own brand of dermatology.

> But if there be doubt about which came first, the white hair or the white patch, The Holy One Blessed Is He says: PURE.
> And the entire membership of the Heavenly Yeshiva (now in Aramaic) all say: IMPURE.

Now it is God, not just Rabbi Eliezer [§21], who is in the minority. What can be done?

> The assembly said: Who can resolve the issue? Rabba son of Nahmani can.
> For Rabba son of Nahmani has said (now in Hebrew): I am the one best expert in the laws of ritual purity.
> They send a messenger after Rabba.

Now, Rabba has two agents on his trail who have been ordered to kill him: the king's agent and Heaven's agent, the Angel of Death.

> But the Angel of Death cannot approach Rabba sitting on the stump, because Rabba does not stop studying Torah even for a moment.
> But the wind comes up, and Rabba hears a rustling among the cane. It must be an ambush set by the king's troops, thought Rabba. And he decided that it would be better for him to die than to fall into the hands of the king.

With no more ado, Rabba is among the assembly in Heaven.

> Upon his death, Rabba said: PURE, PURE.
> And a Heavenly Voice (the daughter of a voice) was heard to say: Oh happy is Rabba son of Nahmani; his body is pure and he has given up his soul in purity.
> A note dropped from Heaven to the Yeshiva in Pumpadita (written in Hebrew): Rabba has been summoned to the Heavenly Yeshiva.

(If you are curious, you might like to know that the *Halakhah* ruling by Maimonides, the great medieval philosopher and talmudist, is in accord with the opinion of the majority, and not in accord with God and Rabba: if in doubt, IMPURE.)

The Rabbis find Rabba's body, and declare public mourning for three days, which a Heavenly Voice extends to seven days. (Rabba apparently is better appreciated in Heaven.) The scene now shifts to another place.

> That day, a Bedouin riding on his camel was suddenly lifted into the air by a whirlwind and man and camel were deposited on the other side of the Papa river.
> What's going on? Says the Bedouin. They explained to him that Rabba son of Nahmani had just died. (Who "they" are, we are not told.)
> The Bedouin addressed Heaven: Master of the Universe, the world is yours, and Rabba son of Nahmani is yours; You are his, and he is Yours. Why do You destroy Your world? The storm rested.

This story of Rabba's death through religious persecution (by whom, the king's agent or the angel of death?) must certainly be unique in religious literature. It goes beyond Akhnai's oven in claiming the right of humankind to interpret a text by its own lights; a text's author, be it God Himself, takes no precedence. We are prepared to admit that Shakespeare cannot tell us the real meaning of Hamlet; every generation lives in its own world and has the right to its own interpretation of Hamlet. But God is not a mere Shakespeare; God stages not only the play, but the theatre and the audience within it, too. The Talmud is very brave indeed.

As a piece of cinematic literature, the death of Rabba is a gem, at least to my mind. The action is dialectical; the composition is contrapuntal. In the first half of the story, the king's agent is out to get Rabba for having organized an Open Yeshiva for the public. In the second half, it is God's agent, the Angel of Death who is after Rabba to join the closed Yeshiva of Heaven. Rabba is wanted in Heaven because he has taken responsibility for expertise in the *Halakhah* relevant to the laws of purity. Heaven apparently is ready to evaluate a person to the degree to which the person dares to evaluate himself or herself. A human is a creature who takes responsibility.

But Rabba evades the Yeshiva of Heaven because he is studying Torah in his own sitting (*yeshiva*; §11) on the palm stump. Rabba only stops studying, and so falls into the hands of the agent of Heaven, because he thinks the king's agents are at hand to take him to the king of Persia; instead he is taken to the King of the Universe.

Note, by the way, that the Talmudic heaven is a copy of the Talmudic earth. No eternal state of spiritual bliss; no union with the One; no garden of sensual pleasure; just continuing cognition: study and more study by the individual in a society. The ideal life is a learning process, both on earth and in Heaven. Heaven, like earth, is not Being, but Becoming.

But the last word is given to a Bedouin. The Bedouin whose plea calms the storm is probably illiterate and, not being Jewish, is technically a pagan; the Talmudic period precedes the founding of Islam by hundreds of years. The Bedouin is not interested in disputations about dermatology or Talmudic interpretations of Torah or *Halakhah*; the Bedouin is the voice of the natural order of things. Let the world go on without the whirlwinds of God; let us go about our business. And God listens to the Bedouin. But our Talmudic Bedouin cannot help being Talmudic: What can he have meant by saying to God that "You are his and he is Yours"? Rabba surely belongs to God, but in what way does God belong to Rabba?

TIME

Subject Three: Time

This section is divided into four parts. Language and time sets the stage by comparing the sense of time inherent in the tense structures of Hebrew and English. Time is central to Judaism. Indeed, God's personal Hebrew name designates a type of existence in time. Varieties of time defines chronological time, natural time, historical time and existential time. Texts of time shows us Judaism's interest in existential time. Holidays of time discusses the concepts of time—historical and existential—inherent in the character of the Jewish holidays.

Language and time

The tense structure of your language affects your sense of time; language creates your world.

§29 Hebrew time

Why is it that only verbs bear tenses? Why don't nouns keep time? Why not mark ideas, objects or persons with past, present or future existence, irrespective of their predicated actions? Perhaps it is only natural that actions—designated by verbs—are the vehicles of time. We conscious beings feel ourselves to be eternal, and so we take it for granted that past, present and future are linked not to us as entities, but to what we have done, are doing, or will do as agents. Events define time, so verbs house tense. Nouns, like we who speak them, are timeless, and so nouns need no tense.

English and other European languages are very exact about the way tense narrates the order of time. The basic tenses are only three: past, present and future. But English, says Edward D. Johnson (*The Handbook of Good English*, Pocket Books, New York, 1991 edition, page 48) has by some counts over thirty tenses. Not only may you say *I wrote, I write, I will write*, you can say *I have written, I had written, I will have written, I was writing, I am writing, I will be writing, I have been writing, I had been writing, I shall have been writing, I do write, I did write, I was going to write*, and on and on. In English you can carve time into very fine pieces by relating the history of one event to another. All the tenses beyond simple past, present and future serve to tell the listener about a chain of events in a particular order. The turkey will have been cooked, I hope, by the time you come for Thanksgiving dinner. Indeed, temporal relationships between events define literal history. The tense structure of English expresses an order of events in time. Hence, we can say that speaking English automatically focuses the attention of speaker and listener on historical time. Not so Hebrew.

In Hebrew you can say *I write, I wrote*, or *I will write*, and that's all. There is no historical fine-tuning of the act of writing within a context of other actions. For example, you cannot in Hebrew express the complex relationship in time signified by an English sentence that says, "for you *to have read* this book, it first *would have had to have been read* by an editor". Hebrew simply does not provide a verbal syntax for relating the time of one action to the precise time of other actions in the flow of events. The simple tenses of Hebrew place each action into its own point of reference. The Hebrew language can be used to express a temporal chain of events, but not in the way English uses tenses.

Hebrew, in fact, might be said to lack a present tense; it could be argued that Hebrew has only two real tenses: past and future. A Hebrew verb in the present tense cannot be distinguished from a noun. The Hebrew words *ani kotev* can be translated accurately into English by saying either *I am writing* or *I am a writer*. Indeed, Hebrew does not even have an equivalent word for the English *am*; the verb *to be* is not used in the Hebrew present tense.

Let us analyze the expression *ani kotev*. The Hebrew *ani* designates *I*; the Hebrew *kotev* designates a *singular male writer*. The Hebrew present tense, if it be a tense at all, has four distinctions: *singular* and *plural masculine* and *singular* and *plural feminine*. The Hebrew present tense expresses the four distinctions of person expressed by Hebrew nouns. The past and future tenses of verbs are conjugated quite differently than is the present tense, and cannot be confused with noun constructions. The Hebrew past and future tenses, however, have their own anomaly, a property unthinkable in English.

Hebrew past and future tense forms can exchange their time reference; a future tense construction can signify the past, and, vice versa, a past tense construction can signify the future. The transformation is done merely by adding the letter *W* to the front of the word. The prefix *W*, or *wa* signifies the English word *and*. In Hebrew, the expression *and I went* can also mean *and I shall go*.

(How then is a reader to know whether a particular phrase is meant to designate a past action or a future action? As you would guess: by *interpreting* the context in which the phrase is imbedded [§9].)

This mutual transformation of past and future is most prevalent in the classical Hebrew of the Bible; it rarely appears in the Hebrew of later times, where it is used in some standard expressions or as a literary device. In Talmudic Hebrew, past is past and future is future, most of the time. Nevertheless, the mutual transformations of past and future are part of the semantic legacy of Hebrew to this day because the Bible still lives in the present language.

We have made this detour into Hebrew to point out that a language provides its speakers with a particular view of time, built-in. The sense of time intrinsic to Hebrew is clearly different from the sense of time intrinsic to English. English time is precisely historical; each event is told in a tense of time that is related to the order of other events. Hebrew time is existential; the person and the present activity of the person are one and the same. The *writer* (noun) is the one who *is writing* (verb), and the one who *is writing* is the *writer*. In Hebrew, time is defined linguistically by reference to the doer. In English, time is defined linguistically by a set of chronological relationships; time is literal history. In Hebrew, the past contains the future and the future contains the past. Time is the conscious will of the doer.

As we discussed earlier, Hebrew reality is linguistic (remember, *davar* is both a *spoken word* and a *thing* [§9]). Indeed, the cadence of sounds that constitute spoken words, as they rise and fall, tick off the progress of time. Hence, the Hebrew grasp of cognitive reality is linked to a verbal sense of time. Greek reality, in contrast, is linked not to time, but to space. Plato connected Being (true reality) to the eternal, unchanging laws of spatial geometry, not to the flow of speech (see *The Republic* and other Dialogues). Hebrew reality is *heard*, Greek is *seen*.

(But is this assertion not contradicted by the facts? The chief Greek prophet, Teiresias, is blind, while Moses, the chief Hebrew prophet, suffers from a speech impediment [Exodus 4:10]. How can the Greek Seer of Truth be blind and the Hebrew Spokesman of God stammer? This paradox, however, only confirms the point: the true Seer, to see the Truth, must be blind to mere appearances and the true Spokesman, to speak the Truth, must be deprived of his own speech. Each narrative points to the essential. Indeed, Phineus the Greek prophet in the *Argonautica* of Apollonias also seems to be blind. The Hebrew prophet Isaiah receives the gift of prophecy—becomes a Moses—when the angel burns the prophet's lips with a live coal [Isaiah 6:6-7]. Elijah, who returns to the Mountain of Moses, finds God not in the wind, not in the earthquake, not in the fire, but in the still small voice [Kings I 19:11-13]. The Rabbis of the Talmud, who supercede the prophets of Israel, also hear from Heaven the *daughter of a voice,* although they may not always accept what it has to say [§21, §28].)

§30 God's personal name

We noted above that one of the general names of God in the Talmud is *Ha-Maqom, The Place* [§19]. Now we are prepared to discuss the most personal of God's names, the name that is related to *Time*.

The generic name for God in the Hebrew Bible is *El,* or a variant of *El, Elohim* (the root is *'L,* which means *power*). The same root appears in the Arabic *Alla,* and in many Hebrew names (for example, *Emmanuelle, God is with us; Raphael, God has healed*). A pagan god, using the same generic root, is called in Hebrew *elil*.

God, in addition to His more general names, has a personal name, a name so personal that no person is allowed to pronounce it aloud (except the high priest who, before the Temple was destroyed 2000 years ago, used to pronounce the personal name of God once a year on Yom Kippur, when he entered the Holy of Holies). God's personal name is designated by four letters: YHWH. The root of that name is H-W-H, which refers to *being,* to *existence,* particularly to *existence in time.* The Hebrew word for present *howeh,* is derived form the same H-W-H root.

A name that is ineffable does not serve well as a name; how do you refer to that which you dare not pronounce? The name YHWH appears many times in scripture; but YHWH is read as if a different word were written in its place. The substitute name is a special variant of the word that means *My Lord,* and that variant is used only to address God in prayer. When not praying, an observant Jew will refer to God simply as *The Name.*

The name YHWH does not appear in the Talmud, except by citation as part of a Torah text, and even then the word is replaced by an abbreviation. I pointed out above that the Talmud is sensitive to individuality and, hence, to names [§17]. God is the individual above and beyond all individuals and His name is beyond all names. We shall be careful not to say God's name, but we can still discuss the meaning inherent in the word YHWH.

Recall the first meeting between God and Moses in the Sinai desert before the burning bush. There God commands Moses to return to Egypt and free Israel from slavery. Moses hesitates, and questions God; when the people ask me "what

PROGRESS

is His name, what shall I say to them?" Moses, I gather, would like to know how God refers to Himself. And God answers "*Eheyeh asher eheyeh*", *I will be that which I will be*. Tell the people, "*I will be*" will free them from the house of bondage (Exodus, 3:14). The Hebrew root of *I will be* is H-W-H, the root of God's personal name. *I will be* is the equivalent of YHWH.

Now consider this, *I will be* relates to two aspects of time that interest us here: One aspect of *I will be* is what we have called *progress* [§2, §30]. *I will be*, as the name of God, expresses a promise for the future. The other aspect of *I will be* expresses the idea of will, of choice. Choice, as we shall discuss below, marks a moment of *existential time* [§38].

So you can see that unique perceptions of reality are inherent in the Hebrew language, and probably in every language. Thinking is talking to yourself, and your language channels your thoughts. The grammatical structure of Hebrew carries with it a sense of time and person that suits the existential moment—the direct transition of future into past. The present in Hebrew is not a tense but a person. The transparency of the roots of Hebrew words triggers semantic associations that resonate even to the throne of God on High. God's names are Place (MAQOM) and Time (YHWH) and Choice (I WILL BE).

Varieties of time

Time is variable; there are different ways to feel it and to measure it.

§31 Progress

Western science rides forth on the belief that nature can be understood, controlled and reformed for the better. The pursuit of progress, as I said in the Preamble, motivates science [§2].

The word *progress* comes from the Latin *progressus*, a step forward, an advance. To step forward means literally to move in space towards a desired objective. But the concept of progress is not a concept of *space*; metaphorically, progress is a concept of *time*. (Concepts of time are often expressed using metaphors from the geometry of space; or perhaps we should say from the architecture of space). You may progress socially or economically within this world by moving to a better neighborhood or to a new job. But nature herself cannot be improved by moving the world to a more favorable place in space; the world is in the place it always has been; it spins and revolves, but it cannot pack up and move elsewhere. The progress of nature can refer only to changes evolving over *time*. Progress describes a succession of events that satisfies. Progress is the music of time. What is *time*?

§32 Four kinds of time

Time travels in divers paces with divers persons, says Rosalind (W. Shakespeare, *As You Like It*. Act III, Scene ii). But few of us ever pause to define time because we sense it without effort. We experience time by the way it slips through conscious awareness (time's presence), by the changes we see in time's wake (time's past), by hopes and plans (time's future). Time can vary with one's point of view; a person can live in different frames of time in the course of a single day; waiting for the dentist differs from waiting for the Messiah. Here we shall discuss four varieties of time: chronological time, natural time, historical time, and existential time.

§33 Chronological time

Chronological time is easy to define; it's the time measured by clocks. Chronological time proceeds independently of human thought or action. The clock ticks off the same intervals whether or not we look at it. We might argue, in fact, that chronological time goes on even when there is no clock ticking and no person watching. Chronological time can also exist as an abstraction: for example, the time that appears in the fundamental equations of nature formulated by physicists and chemists: gravity, relativity, quantum mechanics, etc. However, the abstract concepts of time (and space/time) developed by Einstein and other theoretical physicists extend to scales and realms of reality outside the world of direct human experience and perception; see Stephen W. Hawking, *A brief history of time*. Bantam Press, London. 1988; Roger Penrose, *The emperor's new mind. Concerning computers, minds, and the laws of physics*. Vantage Books. London, 1990; John A. Wheeler, *A journey into gravity and spacetime*. Scientific American Library, New York. 1990. The time of physicists, as fascinating as it is, will be ignored here; we shall limit our discussion to mundane time as we (and the Talmud) feel it roll by.

§34 Natural time

For most of the generations of human consciousness, we have lived, not in the chronological time of equations or ticking clocks, but in time marked by the recurring cycles of nature. The rising and setting sun defines day and night, and so determines our daily cycles of wakefulness and sleep. Months and tides are naturally defined by the phases of the moon. Years can be seen in the to-and-fro migration of the sun on the horizon. Seasons are noticed by warm and cold, by wet and dry, by span of day and night, and vitally by the death and rebirth of vegetation and animal life. Life is chained to these cycles of nature's time. Creatures wax in the spring and wane in the winter. Nature's clock ticks off intervals of birth and death. The revolving doors of nature create what we shall call natural time.

For one hundred thousand years and more, we humans survived in natural time defined by the seasonal hunt for fish and game and the seasonal gathering of plants. Our development of agriculture, beginning about 10,000 years ago, made us dependent on the growth cycles of domesticated plants and the breeding cycles of livestock. Only the industrial developments of the past few hundred years and the electronic and informational advances of recent decades have allowed us some escape from the cyclical chains of natural time. Yet even today, people in the Northern latitudes often need light therapy in the dark days of winter; the normal chemistry of the human brain is tuned to the sun's cycle of day and night. A lack of sunlight can depress a person's spirit, an organic syndrome called seasonal affective disorder, known by its acronym *SAD*. It is only natural that natural time has molded the ideas of humans about nature and our place in nature.

§35 Two faces of nature

Nature simultaneously expresses both order and chaos; whether you see the order or the chaos is largely a matter of scale. The earth, for example, is perfectly ordered when viewed at the scale of the solar system; the earth, photographed from space, appears to be a smooth bluish ball whose motion around the sun is explainable entirely by the laws of gravity and motion. As you zoom in, you begin to see the diverse shapes of continents, mountain chains, river valleys, and other distinctive features. Zooming in to the scale of human existence, you begin to see the chaos. Daylight and seasons are not uniform, but still are fairly predictable: summer days are longer than are winter days, and summer days are generally warmer than winter days. But nature on the scale of human history and human person is disorderly; life is ruled by the accidents of birth and death. Life's progress has always been chaotic: the plague, the draught, the blight, the falling rock, the angry bear, the treacherous heart, the stock market were never predictable, and often unjust. How do human minds explain the regularities and irregularities of life on earth?

Nature logically exists as one continuum of reality. But most human cultures have personified both the order and the disorder of nature as the varied expression of divine powers, gods. The ordered progressions of nature were associated with the inexorable fates of the gods; the disorderly conduct of nature was associated with the whims of the gods. Each competing god expresses a personal will and favors a particular agenda. Nature was not of one mind: Athena favors Odysseus; Poseidon wants him drowned (see *The Odyssey*, by Homer). Mythology was natural; or better, nature was mythological.

§36 Mythology: The wheel of natural time

Before the world was colonized by Western ideology, most human cultures saw the recurrent life-and-death cycles of nature as reflections of the cyclical sagas of the gods; the gods were fickle, but they too were ruled by pitiless fate. Nature was our provider and our destroyer, our blessing and our bane. The death and resurrection of light, warmth, vegetation and animals were manifestations of the recurrent death and resurrection of gods. Consider Adonis, Tammuz, Osiris, Attis, Mithra, Siva, and their consorts and Great Mothers, and their manifestations in human cultures worldwide (see *The Golden Bough. A Study in Magic and Religion*, by J. G. Frazer, cited above [§22]).

(The semiotics here are interesting; one may wonder who or what is the signifier and who or what is the signified: does the cyclical birth/death of the gods signify the cycling of natural time, or does the eternal cycle of nature signify the eternal drama of the gods?)

Scholars such as Mircea Eliade (*The Myth of the Eternal Return, or Cosmos and History*, Princeton University Press, Princeton, 1954), Joseph Campbell (*The Masks of God: Primitive Mythology*, Penguin Books, New York, 1969), Daniel Boorstin (*The Discoverers*, Penguin Books, New York, 1986) and Thomas Cahill (*The Gifts of the Jews*, Nan A. Talese/Anchor Books, New York, 1999) have taught that non-Western cultures view time as an ever-revolving wheel, an endless cycle manifested as the recurring death and rebirth of nature. The wheel of time is inherent in a worldview that perceives all of nature, gods and humans included, to be manifestations of an eternally cycling divinity. All are entrapped in the cycle of time; time's revolving wheel fates all to be crushed and resurrected, killed and reincarnated.

We ought not deride the wheel of time and the cycling gods of mythology as the superstitions of naïve primitives. Humans were perceptive and reasonable realists, even before the rise of the West. The deification of natural time was a conclusion based on the facts of life. Nature cycled, so the gods of nature too were fated to cycle. The mythical wheel of time was not an ignorant superstition; the myth was a metaphor invented (or discovered) to explain reality.

Human values spring from human world-views. Given the vagary and injustice of existence, a wise person can only bow before the inevitable; a reasonable person will never meddle with nature. Icarus dared fly up to the heavens and crashed down to the depths of the sea; Oedipus presumed to change his fate and brought general ruin. Hubris was the original deadly sin. The lesson was clear: a wise individual will keep a low profile, accept fate and merge with nature. Or if you must be an activist, you might worship nature and engage in rituals designed to propitiate, assist, and participate in the mythic drama of cycling nature. Time and fate roll on unperturbed. It was both foolish and impious to tamper with nature; experimentation was off limits. Nature, divinely powerful and unpredictable from day-to-day, could not be improved; progress in natural time was unthinkable. No progress, no science.

Of course, all human cultures exploit unnatural artifacts such as fire, clothing, weapons, music and medications, and many cultures have mastered agriculture, animal husbandry, metallurgy, and architecture. But these technologies are usually considered part of the traditional endowment of the collective; they are gifts of gods or heroes. Western culture seems unique in allowing individuals to defy tradition and invent and innovate. Progress, along with individuality, is born in historical time.

§37 Historical time

Historical time is recorded by a sequence of events, not by equations, ticking clocks or the to-and-fro of sun and season. Historical time is time marked by acts done by humans, or by acts with consequences for the stories of humans. An historical *event* is an act that humans want to remember and tell in word, in script, or in art. Historical time is time in narrative. The flying arrow of historical time can take us to new and better places. Historical time implies progress.

Historical time contrasts with natural time in two essential aspects: geometry and personification. The geometry of natural time is circular; what goes around comes around—again and again. The geometry of historical time is linear; the sequence of history is not periodical—no event is ever an exact replica of any other event. The present is always unique because the past is always progressing. Scholars complain that politicians make decisions without recourse to the lessons of past history. The politicians should not repent; history never repeats itself anyway. This moment is like no other moment before or after it. Historical time is a flying arrow; natural time is a revolving wheel. The personification of natural time is expressed as the fate of the gods, their eternal rebirth and re-death. The personification of historical time is human action.

Historical time should not be confused with the historical record. People can write history very well without accepting the concept of historical time. History is a form of literature that records or interprets events. Historical time is not mere history; historical time is an ideology. Historical time, in contrast to the historical record, arises from the belief that human actions can improve the world.

§38 Existential time

Existential time, the fourth variety of time, differs essentially from both natural time and historical time. Natural time and historical time are public events; existential time is time made personal. Existential time is the time we create by what we do *now*. Existential time is the time before us, the time that invites action, the time of personal events. Existential time is the time within our grasp. It is not the time *on* our hands; it is the time *in* our hands. Existential time is the time of choice and decision. (But once done and past, subjective existential time becomes objective history—existential time is continuously transformed into historical time.)

As observers, existential time is cinematic; it is formed by the *montage* of a chain of singular events—single shots. Time is measured by the importance of each cut and by the shifts between scenes. The time serves the story; a chronological two-hour movie can expand a momentary event or it can contract an eon. Novels may be situated in either historical time or existential time; the cinema is always in existential time.

In terms of time's architecture, chronological time is linear, natural time is cyclical, and historical time is linear. Existential time, in contrast, is neither a line nor a curve; existential time is a point in the now. Existential time may remember a past and it may anticipate a future, but it lives in the fleeting present.

(Note that the English word *present* bears an existential etymology; *present* is derived from the Latin *prae- before* and *sens*, the present participle of *sum*, I am. Thus, the present is that which is before me as I exist—fleeting time.)

Historical time, as we noted above, is an unfolding story, a narrative. Once the story is told, it becomes public record. Historical time belongs to collectives. Existential time, in contrast, is measured privately. Existential time is not the story; existential time is the storyteller, the narrator.

With this background, we can prepare ourselves for the view of time inherent in the Talmud: Cahill thinks that the Torah initiated a transformation of mythical time into the concept of a progressive historical time (see Thomas Cahill, *The Gifts of the Jews*; cited above [§36]). The Talmud, as we shall soon see, completes this evolution, and then goes on to further transform the historical time of the collective into the existential time of the individual. The written medium of the Talmud is existential, and the message of the Talmud is existential: The Talmud teaches us about existential time using a cinema-like, existential format.

Texts of time

I interpret Talmudic texts related to time, existential time in particular.

§39 Shma: Time begins in darkness

Traditionally, one can begin to study the Talmud at any point in the text. Nevertheless, the volumes of the Talmud do have a traditional order; there is a first volume, and it is called Brakhot (below we shall discuss what the word means [§43]). And the first word of Brakhot, which is the first word of the Talmud, is a question, and that question is about time: *When?* The Talmud opens with a discussion about the time at which one is obliged to recite the verse from Deuteronomy, 6:4 that begins with the Hebrew word "*Shma*". The *Shma* verse is central to Judaism, and observant Jews recite the verse twice a day, morning and evening at appointed times (events at appointed times create ceremonies). The Talmud (Brakhot, 2a) puts the question this way:

> When is the beginning of the time to recite the "Shma" in the evenings?

The *Shma* is the first word of a statement about YHWH. The word *Shma* is a command to *listen*,

THE DAY BEGINS IN DARKNESS

2003   between heaven and earth

or better to *understand* [§9]. The Torah verse (Deuteronomy, 6:4) is as follows:

> Understand, Israel, YHWH is our God, YHWH is one.

This translation, I believe, is more faithful to the meaning of the Hebrew text than is the standard English translation found in the prayer book:

> Hear, Oh Israel, the Lord our God, the Lord is one.

However you might translate the verse, the *Shma* declares that there is only one God, and He is YHWH. Monotheism is a core principle of Judaism, and so the *Shma* is recited at moments of extremity as well as twice daily. The Talmud tells that Rabbi Akiva, who we met above as a champion of life [§16] and as a man of flesh and blood [§25], recited the *Shma* verse as he was being tortured to death by the Romans; as he died, the final word *one* issued from his lips (Brakhot, 61b). Jewish martyrs at the stake or in the gas chambers, have recited the *Shma* at the last moment of life.

The Talmud, in its narrative style, does not explain what the *Shma* means; the meaning of the *Shma* is left to individual understanding. The declaration of monotheism probably means different things to different people, and it can mean different things to each person during the different phases of the person's life. But in discussing *when* the *Shma* is to be said, the Talmud implicitly extends the significance of the verse beyond its literal meaning. In linking the recitation of the *Shma* to a phase of time, the Talmud connects aspects of God to the hours of human action. Our task is to read the metaphors, to uncover the covert subtext formed by the questions and answers of the overt Talmud text. The reader, as usual, has to interpret both the text and the subtext.

The overt text is too long and complicated to quote in its entirety, so I shall only summarize my reading of the Talmudic subtext.

The Talmud asks why the Rabbis first consider the evening declaration of the *Shma*; why not begin the discussion by considering the morning recitation (Brakhot, 2a)? The Talmud offers two reasons for considering the evening before the morning: the first reason is textual, the second historical.

(Why, we may ask in the style of the Talmud, are *two* reasons given? Does not one *good* reason suffice? The greater the number of excuses offered by the tardy student, the greater the disbelief of the schoolteacher, or so I recall.)

The text of the *Shma* in Deuteronomy, 6:7, continues with the directive to recite "these words...[the *Shma*]...when you lie down and when you rise". When you "lie down" implies going to sleep, and when you "rise" implies awakening, and so the Torah itself positions the evening *Shma* before the morning *Shma*. Hence, the Talmud is only following the textual lead of the Torah in first discussing the evening *Shma*.

The historical explanation is based on the order of creation recorded in a different section of the Torah. Genesis, 1:5, states, "and it was evening and it was morning—day one"; darkness preceded light in the act of creation. Time itself began in the evening, so recitation of the *Shma* too begins in the evening. The Talmud, I think, implies that the essence of creation is the transition from darkness to light: progress. Time begins in darkness (evening; chaos) and progresses to light (morning; order). So too, one's declaration of the name of God begins at the onset of darkness and then proceeds to a declaration at the onset of light. The association of the *Shma* with the Creation is an association with historical time; time and *Shma* begin with the Creation, in darkness.

§40 One face of nature

Note that the belief in One God nourishes the belief that nature is rational; a declaration of monotheism is an assertion that the world was created by a Creator responsible for all that happens. As we discussed above, a mythology of many gods explains the irrationality of life by the whims of divided powers [§35]. Monotheism, in contrast to mythology, denies the irrationality of a chaotic world; if nature does behave badly, it is probably our fault [§56]. It is also possible that what looks like nature's misbehavior is not true chaos, but only our inability to grasp the rationality of God [§60]. Because there is only One God, nature can have only one face, not two. Recall that the singularity of Adam expresses the singularity of God [§14]. Indeed, a belief in the intrinsic rationality of nature is yet another debt of science to monotheism.

§41 Associations

Now to the time question: at what exact time does the obligation to recite the evening *Shma* begin? The evening *Shma* may be recited at any time from then on till midnight, or, in a pinch, even till the first light that heralds the coming dawn. The discussion here relates to the existential moment from which time forward saying the *Shma* verse will fulfill the commandment given in Deuteronomy, 6:7. (Of course, one may recite the *Shma* whenever he or she is so moved, but saying the verse outside of the prescribed times will not fulfill the commandment to say it.)

Five Rabbis each propose a different activity as fitting to mark the beginning of the time for the evening *Shma*. Considering the many hours available from sunset to morning, the fine distinctions made by the Rabbis for the beginning moment are all the more meaningful. The associations are the message.

1. Rabbi Yehoshua, who represented the majority in the controversy over Akhnai's oven [§21], associates the evening *Shma* with the time at nightfall at which a ritually impure priest may regain a state of purification.
2. Rabbi Meir, who we met crossing the river after the fetching devil-woman [§26], associates the *Shma* with the beginning of the ritual baptism of the impure priest that makes him pure.
3. Rabbi Hanina proposes that the time of the *Shma* is the time the poor man finally ends his daily labors to eat an evening meal of bread and salt.
4. Rabbi Aha relates the declaration of the *Shma* to the time at which most people, not just the poor, return home from work to dinner.
5. Rabbi Eliezer, who was supported by the Heavenly Voice at the controversy over Akhnai's oven [§21], fixes the evening *Shma* of everyday to the same time at which the Sabbath begins on Friday night.

The Talmud questions the differences between these five opinions, but, perhaps with tongue in cheek, asks only which of them is earlier or later chronologically; the Talmud leaves to us to interpret the human, philosophical, or ethical lessons inherent in each of the five associations of time.

Rabbi Yehoshua and Rabbi Meir connect the evening *Shma* with the ritual purification of an impure priest. A priest was allowed to serve in the Temple only when he was in a state of purity; a priest who touched an impure object could be reinstated by a ritual baptism at sunset (see Leviticus, 22:6-7). Ritual purity, you may recall, served as the test case both in Akhnai's oven [§21] and in Rabba's call to the Yeshiva of Heaven [§28]. Yehoshua connects the *Shma* with the conclusion of the process of purification and Meir connects it with the start of the purification. Both see the *Shma* as an act by which a person transforms, or begins to transform, his or her impurity to the pure service of God. According to this metaphor of time, the *Shma* creates a ritual that purifies every person as if he or she were a priest. Indeed, the destruction of the Temple in Jerusalem and the end of the duties of the official priests two thousand years ago have made it possible for each person to become, as it were, a holy Temple; reciting the *Shma* is the key to his or her priestly service. The association of the *Shma* with the Temple ritual frames individual existence within historical time. Individual existence is a ritual of purification.

The actual difference in clock time between the start of the purification ritual and its termination is probably negligible; the Rabbis, I would say, really differ in their view of the *Shma* as the initiation of purification or as the completion of purification. In either case, the purification ceremony is only a passport into the realm of pure service that must follow.

Rabbi Hanina and Rabbi Aha are more current; they associate the *Shma* with people as people, not as surrogate priests. Hanina sees the *Shma* as an act of identification with the poor, the unfortunate of the earth. YHWH is with the poor; the *Shma* is a ritual, not of the Temple, but of the human condition. Note that the poor man is identified with his time of rest, his exit from the day's labor. A human is not a machine and humanness is peace, time-off from the struggle for survival; the recognition of

God is associated with the moments of deliverance free of the employer.

Rabbi Aha is perhaps more profound. The *Shma* is not a matter of priests or the downtrodden; the time of the *Shma* is the time when most people come home to eat with the family. Daily life itself is a ritual in the name of God. Both Hanina and Aha place the *Shma* in existential time; there is no essential difference between them in chronological time.

Rabbi Eliezer, the purist [§21], sees the recitation of the *Shma* as equivalent to an inauguration of the Sabbath. We shall see below [§45] that the two versions of the Ten Commandments associate the cycle of the Sabbath both with the creation of the world and with the exodus of a people from bondage (compare Exodus, 20:11 with Deuteronomy, 5:15). Thus Eliezer connects worlds, divine and human, through his association of the time of the *Shma* with the onset of the Sabbath. Eliezer's association is both historical and existential.

In summary, the Talmud associates the evening *Shma*, a core principle of Judaism, with the Temple ritual, with the poor of the earth, with mundane life, and with the transition from secular struggle to spiritual Sabbath rest. The enthronement of God in human time is all of these processes, historical and existential. In other words, the Talmud fixed the *Shma* as a twice-daily signifier; but what the *Shma* signifies is the responsibility of each person at his or her moment of recitation.

§42 Love thy neighbor

But the *Shma* is not the only core principle of Judaism. The Talmud in Shabbat, 31a, tells of a gentile who asked the sage Hillel to teach him the entire Torah, the essence of Judaism, while "standing on one foot". Hillel could have recited the *Shma*, but instead he said to the gentile;

> That which is hateful when done to you, do not do to your fellow; this is the essence of the Torah. All the rest is interpretation. Now go and study.

Hillel does not mention the idea of the One God; instead he summarizes Judaism through an interpretation of the command to "love your neighbor as yourself" (Leviticus 19:18). The point for our present discussion is that Hillel associates the essence of the Torah with a type of human behavior (or better, with the rejection of a type of behavior unfitting for people). Similarly, the Talmud associates the recitation of the *Shma*, the essential statement about YHWH, with the times of human events. Human action is the measure.

Hillel's message is clear; but why did Hillel not quote to the gentile the text of Leviticus, 19:18, as it is written—"love your neighbor as yourself"? The Torah commands us to bestow on the other that which we love for ourselves. Why did Hillel turn the verse around and forbid us to impose on the other that which is hateful to us? It turns out that the Talmud uses the explicit command to "love your neighbor" in another context, a context that I first found quite shocking.

Sanhedrin, 45a, discusses how the death sentence is administered to a person found guilty of a capital crime; the court has interrogated the witnesses [§14] and has reached its decision: the person is condemned to die, in this case by stoning. The discussion revolves around the question of whether the condemned is to be executed dressed or not. At this point the Talmud states:

> Rabbi Nahman said in the name of Rabbah son of Abuha; the Torah said, "you must love your neighbor as yourself" (Leviticus 19:18), which means choose for the condemned a good death......

The Talmud then states that the *halakhah* ruling about clothing the condemned hinges on whether the condemned would prefer the preservation of his or her dignity (and die clothed) or a faster death (and die naked). Each alternative has its supporters.

The Talmud goes on to describe the actual execution: The Torah mentions death by stoning (Deuteronomy 17:1, for example), but the Rabbis want to ensure that the condemned is already dead, or at least insensible before any stone is cast. So the Talmud has the condemned first thrown to the ground from a platform two stories high. But, objects a Rabbi, a person can die from a fall from a lesser height; why two stories? The Talmud answers:

> The Torah said, "you must love your neighbor as yourself" (Leviticus 19:18), which means choose for the condemned a good death
> Well, if that's the case, then why only two stories—a higher fall will be even more effective?
> Answers the Talmud; a higher fall would disgrace the body of the condemned.

A person's integrity is to be respected even after death; the impact from a fall from too great a height could break open the body.

How bizarre, if not sacrilegious to direct your love of humanity to the execution of a condemned criminal. What does love have to do with death row? Does the Talmud not know the simple meaning of human love? I must admit that Sanhedrin, 45a, made me feel very uneasy, even disappointed with the Rabbis. But on reflection, it became clear to me that the Talmud knew quite a bit about true love of humanity— quite a bit more than do most of us.

True love is an end in itself. True love can only be love without recompense. If our love is used for gain, then our love is merely a means to some other end. No, true love must be love free of reward.

Now love directed specifically to one's beloved spouse, to one's own child, to an honored parent or to a dear friend may reflect an all-embracing love of humanity, but it need not do so. Indeed, how would you feel if your partner in life would love you with the depth of feeling he or she bestows on women or men in general? Love of beauty, virtue, honor, duty or reward certainly does not express love for the essential humanity of a fellow human. Such love is not the essence of love, but only the use of love. Who, I now understood, could be a more apt target of essential human love than the condemned? We don't know the person; we owe them nothing; they will owe us nothing (they will not survive to repay us or hold us in esteem); we do not approve of them or their actions (they have been judged guilty of capital offense); the community will not honor us in this matter. Thus, the Talmudic situation described in Sanhedrin, 45a, however distasteful, is a true test of essential human love; better, it is a flawless demonstration of love precisely because it is distasteful. This is the concrete lesson of the Talmud; love is self-identification with a fellow human when there is nothing to be gained, and nothing extraneous to identify with or admire other than his or her humanness—the humanity you both share.

After all, we too are condemned to die. What do we want for ourselves as we part but some remnant of dignity and less pain? Let us then bestow this pittance of self-concern on our neighbor, who is in this one matter, at least, together with us. So why then do so many die in pain and indignity inflicted by the humanity of intensive or institutional care? Unfortunately, Rabbi Nahman is already gone and can give us no advice. Hillel, however, has already given us good advice; don't prolong another's death in indignity and pain; we don't want that for ourselves. Besides,

says Hillel, "All the rest is interpretation. Now go and study".

§43 Blessings acknowledge time

The essence of existential time is the awareness of the moment. We can grasp the flow of time when we pause to acknowledge time's existence. Existential time is intensified when we stop to place ourselves outside of time's flow. The Talmud creates, as it were, a moment beyond time when it obliges us to bless actions and events. The ritual of blessing marks the moment.

Every event merits its special Hebrew blessing. Events to be blessed include, for example, reciting the *Shma*, reading a section from the Torah, studying, eating various foods, smelling a fragrance, awakening in the morning, inaugurating a holiday, terminating a holiday, marrying, hearing good news, hearing bad news, seeing the wonders of nature, praying and carrying out a religious duty. An observant Jew is obliged to make at least 100 such blessings every day. What is a blessing?

Every Hebrew blessing uses a common formula. All blessings begin with a salute to God:

Barukh ata YHWH...

The first word in the formula is *barukh*, which is usually translated in English as *blessed*. So the salute is said to begin:

Blessed are you YHWH...

But let's take a closer look the meaning of the word *barukh*. The English word *benediction* means to *wish well* (from the Latin *bene*, *well* and *dicere*, to *say*); the English word *blessing* means to *declare something holy* (The Oxford English Dictionary says that the origin of the word *blessing* is related to *blood* as a sign of holiness). But the

Blessing

Hebrew *barukh* means neither good nor holy. The word *barukh* expresses the root *B-R-KH*. That root arises from the word *berekh*, which means *knee*. A *brakha*, the Hebrew word for a *blessing*, is not an act of speaking or wishing; a *brakha* is taking pause to bend the knee, to *kneel*. A *brakha* is a concrete posture of acknowledgement. A *brakha* is a moment of awareness, a conscious pause in the flow of existence. The Hebrew concept of *brakha* creates a boundary in time, just as the Hebrew concept of *qadosh*, holy, creates a boundary in space [§20].

(Parenthetically, we might note that *brakha* contrasts with *Halakhah* in the way that *kneeling* contrasts with *walking*. *Halakhah*, the codebook of Jewish conduct, is the flow of action, walking the world, as it were [§9]. *Brakha* is stopping the flow of time to wonder about the world, to choose. A *brakha* highlights a shift in scene.)

So each Hebrew blessing begins with an acknowledgement of God by a pause in the flow of time. We ought to translate the standard form of blessing not as

Blessed are you YHWH...

but rather as

Acknowledged are you YHWH...

Note another feature of the *brakha* formula; the sentence shifts its grammatical person. Every *brakha* begins by addressing YHWH in the second person—you. The formula then switches abruptly to the third person; the *brakha* continues

Acknowledged are you YHWH, our God, King of the universe who has commanded..."

"King of the universe who has commanded" is in the third person—*he* who *is* king, not *you* who *are* king. Thus the individual addresses YHWH, the personal name of God, in the familiar second person; the attributes of *God, King of the universe,* are then described in the remainder of the blessing at the distance of a third person. Speaking to God is a dynamic process—YHWH is present before you, His attributes and commandments are more distant [§60]. The second person *you* designates a dialogue; the third person *he* introduces a description. The *brakha* is both a meeting *with* God and a statement *about* God.

The tense structure of the *brakha* also bears meaning. Blessings that acknowledge gifts of God describe the particular act of giving in the present tense. For example, the blessing over bread is "Acknowledged are you YHWH...who *brings* forth bread from the earth"; the blessing over wine is "...who *creates* the fruit of the vine"; the blessing that inaugurates the Sabbath is "...who *sanctifies* the Sabbath"; the blessing on reading the Torah is "...who *gives* the Torah". Although the Sabbath and the Torah are both perceived to have been granted by an act of God in the historical past, the blessing for each uses the present tense.

As we noted above, the Hebrew present tense is indistinguishable from a noun [§29]. If we consider the present tense as equivalent to a noun, then we might conclude that using the present tense to describe an act of God is like describing an attribute of God Himself. The Talmud, you remember, calls God *maqom*; "He is the place of the world" (Bereshit Rabba, 68); the whole world is an attribute of God [§19].

However, we can also view the present tense as a verb signifying an action; in this light, a blessing in the present tense declares that the workings of God in the world are continuously present. Creation is not a terminal activity; the created exists through a continuous act of creation.

In contrast to the use of the present tense to describe God's gifts, blessings that precede one's fulfillment of a commandment use the historical past tense. For example, the blessing made before lighting the Sabbath candles is "Acknowledged are you...who *commanded* us to kindle the Sabbath lamp". After the blessing, the person performs the commandment and lights the candles. The past-tense blessing thus connects a person's present act to a past commandment of God. The *brakha* isolates the existential moment of fulfillment and links it to an unfolding history.

§44 Blessing time

Observant Jews acknowledge not only events in time, Jews bless time itself. The Talmud prescribes the formula for blessing time in Brakhot, 37b, which may be translated like this:

> Acknowledged are you YHWH, our God and King of the universe who has kept us alive, and sustained us and made us reach this time.

This blessing over time is said at moments of renewal, moments of anticipation and celebration: when we inaugurate a holiday, when we marry, when we meet a close friend we have not seen for more than 30 days, when we acquire a new personal possession, when we taste a season's first fruit, when we hear good news, when we attain a goal. In short, we pause to recite the blessing of time at the fulfillment of a longing.

Note the tense of the blessing made over a gift of time:

> "...who has made us reach this time".

God's grant of time is acknowledged here in the past tense. In other words, the *brakha* for the gift of time is not a present-tense acknowledgement of God's continuing creation; one acknowledges

God's grant of time using the past tense structure of a *brakha* said before the fulfillment of a commandment. Is the fulfillment of time a gift of God to us, or is the fulfillment of time a commandment, our gift to God? Perhaps it is both.

Holidays of time

The Torah replaced the ever-cycling natural time of mythology with the progress of historical time ruled by One God. The Talmud proceeds to transform the historical message of the Biblical holidays into existential time. Existential time is the time of the individual.

§45 Replacing nature

We can see a transition from natural time to historical time in the way the Torah remodels the universal celebrations of the agricultural year. In ancient Israel, human sustenance depended largely on a natural cycle that recurs yearly: The winter is the time of the rains, when people sow the grain; the spring heralds the ripening of the wheat and rye at the end of the rains; the early summer is the time of the grain harvest; and the autumn marks the harvest of the summer fruits—grapes, dates, pomegranates—and the return of the rains. These seasons characterize the cycle of natural time in the Middle East, and for ten thousand years people have celebrated this yearly cycle with rites of fertility and resurrection (see, James G. Frazer, *The Golden Bough. A Study in Magic and Religion* cited above [§22]).

Biblical Judaism adopted the three natural holidays of the yearly cycle of spring, summer and autumn, but reinterpreted their meaning. The three holidays remained fixed in natural time as signifiers; the Torah only rewrote their signification. Natural time became historical time. The

Torah does not connect the natural cycle of the grain crop with the recurrent death and rebirth of a handsome young god; instead, the spring Passover holiday commemorates the singular moment of the Exodus of a people from slavery [§46]. The holiday of the harvest and the hope for rain at the end of the dry summer becomes transformed into Tabernacles (Sukkot), a holiday that commemorates the wandering of a people in the desert on their way to a promised land [§63]. Only the holiday of the grain harvest, Pentecost, remained in the Torah without a clear association to history. The Talmud, as we shall discuss below, gives Pentecost an historical connection and so completes the transformation of natural time into historical time, to which the Rabbis then added existential dimensions [§51].

The Torah presents the sacred history of the people of Israel, and at the same time erases the mythological significance of the cycles of nature. Torah history does not live within the mythological wheel; Torah history replaces the wheel. The events recorded in the Bible are always unique; they are chapters in the unfolding of a linear narrative. The story goes on. The Exodus from Egypt leads on to the wandering in the dessert, the settlement of the Land of Israel, the Judges, the Kings, the Temple, and on and on.

Other cultures too wrote history, but only the Jews replaced the mythic narrative of the gods by the ideology of an historical narrative of human development through dialogue with One God. True, the yearly cycles of the holidays persist in Judaism, but these holidays are cycles of human memory and not cycles of nature. The Bible demystified the cycles of nature, and at the same time, established a mystical cycle of its own invention—the week.

§46 Sabbath cycle

The only connection of the Sabbath week to nature is the counting of days: six days of unrestricted activity and a seventh day of rest—sunset to sunset. The week has no reality in any cycle of sun, moon or season. Where does the week come from? Each of the two versions of the Ten Commandments refers to a different historical event to explain the seven-day cycle of the Sabbath rest. (Again, we get two reasons.) Exodus, 20:11 recalls the creation of the world: God created the world in six days and rested on the seventh day. "Therefore, God set aside the Sabbath day and made it holy". The Hebrew word *shabbat* comes from the root SH-B-T—to desist from physical activity. God completed creation and so created the Sabbath day. The Sabbath cycle, according to this version of the Fourth Commandment, signifies not nature, but the creation of nature.

The second version of the Ten Commandments appears in Deuteronomy 5: 6-18. There, the Fourth Commandment to rest on the Sabbath day connects the individual to the collective history of bondage in Egypt. The memory of slavery and redemption obligates each person to observe a ritual day of rest and to provide a day of rest for all others, including servants, beasts of burden and the poor stranger. The Sabbath cycle, according to this version of the Ten Commandments, is a ritual of the human condition, not a ritual of creation. The Sabbath defines time by collective memory and individual responsibility. The practical Roman world, in contrast to the Talmudic world, rejected the cycle of Sabbath rest as an unnatural prize for the lazy and an affront to social rank. Different views of time support different worldviews and ethical values.

§47 Passover and significance

The Torah, as we said [§45], transformed the prehistoric spring holiday of the ripening grain into the historic holiday of Passover, a commemoration of the Exodus of Israel from slavery in Egypt that took place over three thousand years ago. The Passover holiday begins on the evening of the full moon in the first month of the spring. (A new day, you will recall, always begins in darkness; §39.) The holiday lasts for seven days; the first and last days are major days of celebration. Passover is rich in symbols signifying the historical events of the Exodus.

The unleavened bread (the *matzah*) is one of its best-known symbols of Passover. The Talmudic text recited on the first evening of the holiday, the *Haggadah* (which we shall shortly discuss), provides two different explanations for the commandment to eat unleavened bread on Passover: we eat *matzah* on Passover to remind us of the poor food that the slaves, our ancestors, ate in Egypt, and we eat the *matzah* because it recalls the haste of the Exodus; the deliverance wrought by God was so sudden that there was not sufficient time for the redeemed to prepare leavened bread for the journey to freedom. It seems to bother the Rabbis not a whit that a single signifier, the *matzah*, might signify mutually contradictory concepts: let's get the story straight; is *matzah* slave food, or is *matzah* redemption food? Or is it both together? Or do the Rabbis only want us to think about the paradox?

Matzah reminds me of my first visit to the Sinai desert in 1968, before there were any decent roads and the Sinai had not yet become accessible to tourists. On the first night we camped out under the desert sky, our guide, a Bedouin from the Jebeliya tribe prepared bread from dough baked on a metal plate heated over our open fire: the bread was identical to the Passover *matzah*. Is this too what the Passover *matzah* signifies, the bread eaten by Israel wandering in the desert?

One could claim that certain rituals of the Passover holiday might have entered Judaism (or might have been retained by the Torah) from the natural and mythological holidays that preceded the Exodus: does the *matzah* antedate Israel, a remnant of the natural holiday of the ripening grain? Did bio-technology begin with the discovery that yeast could be used to make dough rise, or did the discovery of the fermentation of beer or wine precede bread? Perhaps the *matzah* is eaten to recall the simplicity that reigned before humans learned to rule nature?

Other Passover symbols, especially those connected to the Paschal lamb and the death of the Egyptian first born may be related to ancient rites of resurrection and redemption [§36]. It is no accident that the Christian Easter is connected to Passover and that both celebrations are connected to rites of spring.

Indeed, Christianity transformed the Passover meal into the "last supper" of Jesus and the *matzah* into the consecrated bread that becomes (or signifies) for the Christian the flesh of Jesus. The idea that *matzah* can be transubstantiated into flesh may have contributed to the grotesque anti-Semitic fantasy that the Passover *matzah* is prepared with the blood of a ritually murdered child, Christian or Muslim depending on the locality. This blood libel has burdened the *matzah* with yet another layer of symbolism—that of murderous anti-Semitism. What other food can rival the semiotic reincarnations undergone by poor unleavened bread?

Let us return to Passover; why doesn't the Talmud give us a single, definitive explanation for eating *matzah*? By providing contradictory explanations, the Rabbis may have intended to teach us that the *matzah* bears no authorized meaning. What is a signifier anyway? Like the *Shma* [§41], the matzah is a symbol that can (perhaps, should) signify different ideas to different people at different moments. *Matzah* is only *matzah*, and so each is free to taste in *matzah* his or her own slavery, freedom, desert night, spring day, Savior or family home, or even just the horseradish and garlic. The interpreter must find her or his own significance.

Judaism imposes on the Jew an elaborate code of behavior [the *Halakhah*; §10]; but this *Halakhah* comes without a dogma; there is no authorized interpretation that defines the significance of particular behaviors imposed by the code. Thus we can view the *Halakhah* as an obligatory code of *signifiers* whose *signified* concepts and values are left to individual *signification*. As we discussed above, our ability to recognize a signifier follows from our knowing that the signifier does indeed represent a signified concept [§5]. We don't have to know exactly what the signifier signifies, only that there is something signified. Passover is clearly a signifier; what does it signify?

§48 Telling the *Haggadah*

Passover, as we have discussed, is a holiday of history. Now a holiday is a ritual; how is history transformed into ritual? The ritual meaning of the Exodus, of course, is only a particular instance of a larger question: what does past history mean for one's life in the present? Passover calls attention to the significance of history.

Like all history, the history of the Exodus is embodied in a text. A text of history can be verbal and celebrated in song, poem or saga. The ritual history of Passover exists in two written texts—the Torah text and the Talmud text. The history of Passover is recorded in the Torah in the latter part of Genesis and in the first part of

Exodus. The Torah story is dramatic, detailed and personal; we meet Joseph and his brothers, their father Jacob, Potiphar's wife, Pharaoh, Moses and his brother Aaron and sister Miriam. We read of the burning bush, the confrontations, the plagues, the Exodus, the splitting of the Red Sea. The Torah is historical narrative.

The Talmud text for Passover is called the *haggadah*; the word *haggadah* means *telling*. The *haggadah* is the text read at the Passover eve table by the extended family during a ritual meal called the *seder*. The word *seder* means *order* or *organized ceremony*.

The Torah commandment to celebrate the Passover includes a verse:

> And you will tell your son on that day (when celebrating the Passover) saying, for this did YHWH do for me when I left Egypt (Exodus, 13:8).

So Jews are commanded by the Torah to *tell* the story of the Exodus, and that *telling* is recorded in the *haggadah*, a text of *telling*. Logically we would expect the *haggadah* to recount the history of the Exodus as it is described in the Torah narrative.

§49 Narrative and narrator: Telling and hearing

Despite this expectation, however, the text of the *haggadah* does not repeat the Torah text that tells us the history of the Exodus. The *haggadah* text does not even mention Moses, the man who led Israel out of slavery and into a Covenant with YHWH. So where in the *haggadah* text of the Passover ritual is the history of Passover? Passover, as we discussed above [§47], marks a transformation of the wheel of natural time into the arrow of historical time. History is the narrative of past events; the *haggadah* is supposed to signify the history of the Exodus. So how can the *haggadah* dare ignore the official narrative of the Exodus? What else but the story itself could the *telling* tell? Let's look at the text of the *haggadah* itself. Its message is in its structure.

The *haggadah* includes directions for ritual acts, along with readings to be said aloud. (Reading aloud is important; remember the Hebrew word *davar*, which means both a *spoken word* and a *thing* [§9]; words create [§27]). The seder opens with the *brakha* over the wine that sets the time of Passover apart from the nameless flow of secular time; this inauguration of the holiday is followed by the *brakha* for time itself [§43, §44]. The *haggadah* then goes on to prescribe the performance of various symbolic acts that include ritual hand washing (purification?), blessing the eating of a spring vegetable (inaugurating the spring?), splitting a *matzah* into two pieces (whatever that might signify), and continues with declarations by the father (the leader of the ritual) that the *matzah* is the bread of slavery; that the poor are invited to join our meal; and that we are all in transition from a lower state (slavery, exile, spiritual poverty) to a higher state (freedom, redemption, spiritual progress). Then comes a dialogue in which the son (the young generation) asks the father a series of specific questions regarding the various symbolic peculiarities of the *seder* night, and the father responds by saying that we all have to *tell* the story of the Exodus, the more the better. This is the point where one would expect the *haggadah* to do what father says and proceed to tell the story of the Exodus as it is written in the Torah. But, as I said above, the Torah story is not our text.

Instead, the *haggadah* text goes on to tell us about five Rabbis of the Talmud (including Rabbi Yehoshua [§21, §23], Rabbi Eliezer [§21-22], and Rabbi Akiva [§16, §21-23, §26, §39] whom

we already have met) who themselves are described telling each other the story of Passover. We next read about four hypothetical sons who (like our actual son now sitting at the table) ask different questions about the Passover ritual; we read an analysis of the verse (Exodus, 13:8) that commands us to *tell* the story; we read a synopsis of the history of the Exodus told by Joshua (Joshua, 24: 2-4); we read that God foretold father Abraham that his children would be enslaved in Egypt, and then redeemed (Genesis, 15: 13-14). Finally, the *haggadah* does get around to some citations from the Torah narrative, but the quotes from the Exodus story are couched within a convoluted Talmudic explication of yet another synopsis of the Exodus; this time the narrator is a nameless person who is destined in the future to bring an offering of agricultural produce to the Temple in Jerusalem (Deuteronomy 26: 5-8). Bringing the season's first fruits is accompanied by a formalized statement about the redemption. That statement is used by the *haggadah* as a vehicle for quoting verses from the Torah narrative of the Exodus.

The Haggadah seems here to be turning time on its head; the past Exodus is summarized by the Torah's proclamation put into the mouth of a yet-to-be-born person destined to bring an offering to the yet-to-be-built Temple in Jerusalem.

This expanded synopsis of the Exodus is followed by a series of symbolic rituals, more Rabbinic interpretations of Torah verses, praise of God, a return to the *matzah* and other Passover symbols, a second ritual hand washing, and the meal itself. The *seder* ends with psalms of praise and communal singing.

The structure of the *haggadah* has long perplexed scholars; the *haggadah*, the *telling*, tells all but the historical narrative of the Exodus itself. Instead of telling the story, we tell about

HISTORY

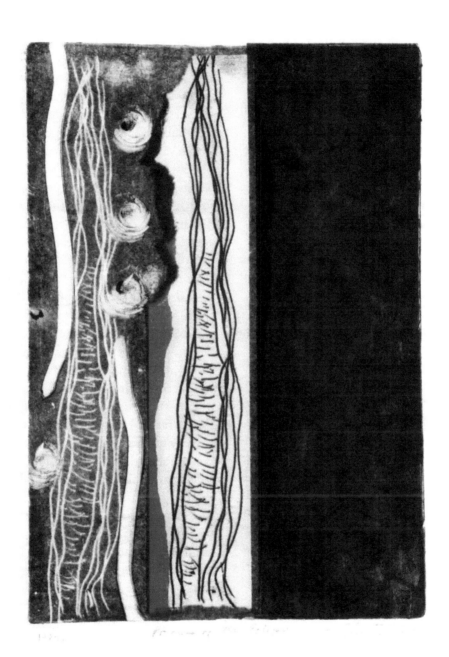

other people who themselves are telling the story. The *haggadah* presents us with a mixed set of *narrators*: the father who leads the ritual, the sons who ask questions about the ritual, five Rabbis of the Talmud who discuss the story, God who foretells the story, Joshua who reports the story, and, finally, a short story told by one who is a future descendant of the redeemed. This last abstract of the story is used by the Rabbis to cite parts of the history of the Torah. Instead of history, the *haggadah* contains a play within a play within a play, or, if you like, a story about various tellers telling about the story. Among the narrators appearing in the *haggadah*, the only one who has had first-hand experience of the Exodus is Joshua—Moses' second in command; even God's revelation of the Exodus to Abraham was a prophecy before the fact. Where is Moses, the instrument if not the hero of the story? What is the *haggadah* telling us by such a labyrinthine construction of what should be a simple story? Has the Talmud overdone Talmudic sophistry?

To my mind, the *haggadah* makes sense if we frame the *haggadah* within a concept of *time*; the *haggadah* is true to the Talmudic concept of existential time. As we discussed above, the essence of existential time is not the *narrative*; the essence of existential time is the *narrator* [§38].

Consider the verse in Exodus (13:8) that commands the *telling* [§48]: "And you will *tell* your son on that day saying, for this did YHWH do for me when I left Egypt". My translation may seem a bit clumsy but only because the Hebrew verse itself is fuzzy. The King James version of the Bible attempts to set the verse right by translating it like this:

> And thou shalt show thy son in that day, saying, This is done because of that which the LORD did unto me when I came forth out of Egypt.

This translation seems quite reasonable; it says that the "this" (the *seder* ritual) is done to commemorate the historical Exodus. But, however reasonable, this translation, at least to some, is wrong.

Rashi [§6], the authoritative rabbinical commentator from Medieval France and Germany (1040-1105) and Ibn Ezra from Spain (1089-1164) both argue that the plain sense of the verse is the very opposite of the interpretation built into the King James translation; the Torah verse explicitly states that the Exodus took place so that we may celebrate our *seder*. In other words, God delivered us from Egypt long ago so that we could perform the *seder* ritual, now. The Exodus took place so that we alive today could make a *seder* and read the *haggadah*. The fact of history, the Exodus, is not the reason for the ritual; no, the performance of the ritual by us at this table on Passover eve is the reason for the historical Exodus. Let me say this in other words; the past is the stage upon which we must perform the acts of the present, at this moment. One's history, indeed the history of the world, creates the theatre in which one must choose his or her act in existential time. Historical time is the narrative. Existential time is the narrator. The ritual is not a symbol of past history; the ritual is symbolic of itself. Think of it this way: The *seder* is not a signifier of the Exodus narrative; the Exodus narrative is a signifier of the seder. The Passover ritual of history is the act of adding ourselves to the long list of narrators who tell about the narrators in the *haggadah*. The *haggadah* is a cinematic script that fashions existential time out of the flash-backs told by a collection of narrators, each narrator tells his version from a different point of view and from a different memoir. We, too, add our own memoirs from *seder* to *seder*.

(Nevertheless, the King James version of Exodus 13:8, to be fair, does rest on authorized support. Nahmanides (1194-1270) of Gerona, Catalonia (now Spain), whose commentary on the Torah is standard alongside that of Rashi, prefers the historical sense of the verse to Rashi's existential sense. Commenting on Exodus 13:8 and especially on 13:16, Nahmanides insists that signifiers such as the *seder* do signify the acts of God in history, and not the other way around as Rashi the existentialist would have us believe. As by now you must know, I prefer Rashi. Rashi preceded Nahmanides by 150 years, but Rashi, at least to my mind, is more current.)

Note that *haggadah*, telling, refers to speaking and hearing, not to reading the written word, but to the saying of it [§9]. The passing of time, of course, is marked by impermanence; things come and things go. Speech exemplifies time; it depends on the rhythm of vanishing words [§29]. The sound of each word must die out to be replaced by the sounds of the coming words. You understand speech by hearing a passing string of words. But just as hearing marks time, seeing marks space. The written word, in contrast to the spoken word, is a pattern in space rather than a pattern in time. The word *haggadah,* then, connotes an existential point of view. True, the text is written in permanent ink, but the *haggadah* is fulfilled in saying the words and hearing them—the telling of it. The *haggadah* is thus expressed in speaking time, not in written space [§9].

§50 Meta-(meta-text)

Let me interrupt our analysis of Talmudic texts for a moment and look critically at my exposition of the *haggadah*: Is it accurate to view the sub-text of the *haggadah* within a framework of existential time, a concept developed almost two thousand years after the canonization of the *haggadah*? Even Rashi and Ibn Ezra lived a thousand years ago; how do I describe their ideas with such post-modern anachronism? Have I invented a sub-text for the *haggadah* that never existed in the minds of the Rabbis who formulated it? How can I say that the Talmud is cinematic; the Yeshiva is not a movie house.

I can respond to this criticism in two ways (which probably means I don't really have the true answer; see §39):

The first answer is philosophical: We here at the *seder* table are not interested in the *history* of the *haggadah* text, but in the *meaning* of the *haggadah* text. What any text means to its reader does not depend on the intention of the author. We may ask Shakespeare what he Shakespeare had in mind when he wrote The Tempest, but Shakespeare is not more expert than are any of his readers, performers or theatre audiences in interpreting what Prospero means. Rabbi Yehoshua, you recall, rejected God's attempt to support Rabbi Eliezer's ruling on Akhnai's oven; a text once given, said Yehoshua, no longer need obey the will of its author, even if the author is THE AUTHOR [§21].

We can also put it this way: Past history is only the stage for the present [§48], so my ideas about the meaning of the *haggadah* stand or fall on their own merit, no matter what the historical Rabbis had in mind. There is no way for us in the 21st Century to know what the Rabbis really thought. We have only their text and our interpretation. The Rabbis used the Torah text to construct their own meta-text, and I have taken the Rabbis' meta-text and have constructed a meta-(meta-text) of my own.

(In fact, no person can ever know what another person, even a spouse, has in mind; we have only the text of their words and our own interpretation. Understanding is translation, the creation of a meta-text; see George Steiner, *After Babel. Aspects of Language and Translation*, Third Edition, Oxford University Press, Oxford, 1998.)

The second answer is historical: Judaism is intrinsically existential, even if the word *existential* is foreign to the classic texts of the culture. Individualism ("for me was the world created" (see §14 and §15-§17), freedom of choice (see §21), the obligation to intervene in the world (see §27) and the association of time with human action (see §41, §43, and §44) are inherently existential. There is simply no clearer way for us to conceptualize these aspects of classic Jewish texts.

We might argue that the Talmud has remained alive for so long in so many different parts of the world precisely because of an enduring modernity; the Talmud is a dynamic process and not a dated precept. The Talmud is a cinematic genre, even if the actual cinema took two millennia to catch up with it. The *haggadah* has remained the most popular of Jewish texts, despite its paradoxes—perhaps because of its paradoxes. Like a good flick, no one can dictate what the text means, neither by reason nor by dogmatic authority. Hence, the *haggadah* lives to generate renewed interpretation, ever-new meaning. The Exodus, the whole of Judaism in fact, is an ongoing *process* of redemption (Becoming), not a fixed *state* of redemption (Being).

§51 Shavuot: And fulfillment

The holiday called Shavuot takes place fifty days after Passover. The word *shavuot* means *weeks* in Hebrew; *shavua* (the singular form of *shavuot*) bears the root of the word for seven, *sheva*.

(It is conceivable that the words *seven* and *sheva* may both derive from an aboriginal stem language that gave rise to Semitic and Indo-European branches.)

The Torah instructs Israel to count daily the ripening grain for seven weeks (*shivah shevuot*; 49 days) starting when the standing stalks can be cut with the scythe. The counting terminates on the 50th day, on Shavuot, when the grain has become fully ripe for harvest (Deuteronomy 16:9-10). Shavuot in English is known as Pentecost, which comes from the Greek word meaning *fifty*.

The Torah transformed, as we discussed [§46], the spring celebration of Passover into a celebration of history unique to Judaism, but the Torah left Shavuot unmodified; Shavuot remained an agricultural ritual similar to the natural holidays of the grain crop celebrated for thousands of years by people the world over (See Frazer, *The Golden Bough. A Study in Magic and Religion* cited above [§22]).

The Talmud, however, completed the historization of time; Shavuot was interpreted by the Rabbis to signify the giving of the Torah at Mount Sinai by God to the People of Israel. The connection between Passover and Pentecost was now matched both in nature and in history: in the cycle of nature, the harvest (celebrated by natural Shavuot) is realized 50 days after the standing stalks of grain first appear (celebrated by natural Passover); in historical time, Israel's Covenant with God at Sinai (the historical Shavuot) fulfills the redemption from slavery that took place 50 days earlier (the historical Passover). Pentecost thus fulfills Passover in two scales of time: The wheat is resurrected from the ground (Passover) to sustain humanity in harvest (Pentecost); Israel rises from human bondage (Passover) to sustain itself in the service of God (Pentecost).

§52 Biblical Shavuot: Receiving

The Rabbis provide two collections of texts to help us interpret the significance of Shavuot: texts from the Bible and texts from the Talmud. The Bible texts are read during the synagogue service of the holiday. Each holiday is marked by specific texts from the Bible selected to be chanted aloud during the morning service in the synagogue. The particular Bible texts chosen for each holiday service constitute a Talmudic meta-text, an implied commentary on the meaning of the day; see Megillah, 2a.

Three different Biblical texts are read on Shavuot. Text one is the Torah narrative that describes the giving of the Ten Commandments at Mount Sinai (Exodus, chapters 19 and 20); text two describes the vision of the Heavenly Host reported by the prophet Ezekiel (Ezekiel 1:1-28, and 3:12); text three is the Book of Ruth (Ruth, chapters 1 through 4). Each text is a masterpiece of Biblical literature; but the texts are poles apart in theme, tone, language, character, person and place. Reading three such diverse texts in the same service highlights their differences; the contrast between them creates a meta-textual commentary on the ways humans may connect to God. Let us compare the three texts:

The Torah text is a volcanic eruption: the narrative is dramatized by smoke, fire, trumpet blasts, and the voice of God Himself. The setting is the holy mountain in the desert. The personae are the people as a collective, Moses their leader, and the public epiphany of God. There is no dialogue; the spoken words are commandments. The emotion is fear and awe. The historical setting is the trek through the desert of the newly freed slaves towards the Promised Land. The people gathered at the foot of the mountain accept the Covenant offered by God: "And the people answered as one, whatever God has spoken we will do" (Genesis 19:8).

Ezekiel connects to God in a different way. In contrast to God's revelation before the nation at Sinai, Ezekiel alone envisions the Heavenly Host of the Lord. Antithetic to the epiphany at Sinai, Israel, at the moment of Ezekiel's vision, is on its way out of the Promised Land; Ezekiel is among the exiles being transported to captivity in Babylon after the conquest of Jerusalem and the destruction of the First Temple by Nebuchadnezzer (see II Kings, 25:1-11). Ezekiel's experience takes place on the banks of a river in Babylon. There are no commandments, only ecstasy and wonder. Ezekiel is the passive vessel of ecstatic revelation and sublime experience; he has no choice.

The story of Ruth is yet different. Here is a short synopsis of the Hebrew narrative:

In the days of the Judges, a man called Elimelech leaves Judea with his wife Naomi and his two sons for the land of Moab, to escape a famine in Israel.

(Recall how the saga of Passover begins when Jacob and his family leave Israel for Egypt also to escape famine; exile is consequential in human events.)

In Moab, the two sons marry local women Ruth and Orpah. But suddenly Elimelech and his sons die leaving three widows: Naomi and her two daughters-in-law. Naomi begs the young widows to stay in Moab to remarry and rebuild their lives, and sets off to return to Judea a poor and beaten woman; she is no longer *Naomi* (*pleasantness*) but *Mara* (*bitterness*; see Ruth 1:20-21).

(The names of the personae in the Book of Ruth are signifiers: as the Talmud says, *Orpah* means to *turn one's back* [Ruth Rabba, 2a]. The word *Ruth* does not bear a clear meaning in Hebrew; it could be related to the words for *friendship* or for *foresight*).

Orpah kisses Naomi and *turns her back* to return to her home in Moab, but Ruth clings to Naomi;

> And Ruth said, Entreat me not to leave thee, or to return from following after thee: for whither thou goest, I will go; and where thou lodgest, I will lodge: thy people shall be my people, and thy God my God: Where thou diest, will I die, and there will I be buried: YHWH do so to me, and more also, if ought but death part thee and me. (Ruth 1:16-17; translation based on the King James Version).

Ruth's devotion to Naomi is Ruth's connection to YHWH; Ruth has received the Torah.

Naomi and Ruth arrive in Bethlehem at the barley harvest (Shavuot). The two women are impoverished, and Ruth joins the poor gleaning the fallen sheaves after the reapers (see Leviticus 23:22). She happens to reach the field of Boaz (*in him is strength*); Boaz is a family relation of the dead husbands of Naomi and Ruth. Ruth finds favor in the eyes of Boaz, and Naomi encourages the relationship by sending Ruth to sleep at the feet of Boaz at night on the threshing floor.

(Strangely, this verse is among the several places in the Bible where there is a divergence between a word as it is written and the word as it is to be read: Naomi's instructions to Ruth, "*you* (Ruth) will go down to the threshing floor" and "*you* will lie at his feet" are written as if, "*I* (Naomi) will go down" and "*I* will lie at his feet" [Ruth 3:3-4]. Although the verb form of the first person "I" is written, the word is to be read as "you". Is this a lapse into an archaic form of Hebrew, or is it a Freudian slip of the quill that precedes Freud by several thousand years?)

Boaz awakens to find Ruth lying next to him. He is in love with Ruth, and wants to marry her. But it turns out that another, closer relative has a prior right to marry Ruth (see Deuteronomy 25:5; levirate marriage). Boaz confronts the other man, and this man chooses to relinquish his right to marry Ruth. Ruth and Boaz marry, and Ruth gives birth to a child, a grandchild for Naomi who restores the pleasantness of her life;

> And the women her neighbors gave it (the child) a name, saying, There is a son born to Naomi (do the neighbor ladies know about Naomi's Freudian slip?); and they called his name Obed: he is the father of Jesse, the father of David (Ruth 4:17).

In contrast to the first two texts, God makes no personal appearance in the Book of Ruth; He does not even seem to guide the action. The events in the first two texts involve the to and fro movements of an entire people between redemption (from Egypt) and exile (to Babylon); the connection of Ruth to the God of Israel involves the exile and redemption of two women, Ruth and Naomi to and from old and new homes. The setting is the pastoral village of Bethlehem, the harvest, and the quiet night among the new-cut sheaves. The sounds are soft human voices. The subjects are women and their men; the story is a woman's story. There are no commandments. The emotions are love and kindness. The essence of the story of Ruth is loyalty, intimacy and choice. The enduring outcome of the union of Naomi and Ruth and of Ruth and Boaz is the dynasty of David, momentous for Judaism, for Christianity, and for our world.

In their choice of Biblical texts, the *montage* of Shavuot, the Rabbis show us three variations on a theme: One can receive the Torah through a collective national commitment (Exodus), through a singular religious experience (Ezekiel), or through a woman building a new home in loving kindness (Ruth).

§53 Talmudic Shavuot: Choosing

The second set of texts interpreting Shavuot are Talmudic texts. Many Talmudic texts explore different aspects of the meaning of Sinai, the connection between Israel and God. Here I will focus on two questions that bothered the Rabbis, and bother Israel to this day: why did God choose to give the Torah to us? Why did we choose to accept it? The Talmud states that God proposed His Covenant to all the nations, but only Israel would have it.

> Rabbi Yohanan said; we can conclude that the Holy One Blessed is He offered it (the Covenant) to each and every nation and tongue and they did not accept it, until He came to Israel, and Israel accepted it (Avodah Zara, 2b).

The story is enlarged in Sifre Brakha (cited in *Sefer Ha'Aggadah*, by Haim N. Bialik and Y. H. Ravnitzky, Devir, Tel Aviv, 1956, page 59). When God offered each nation the Torah, sensibly each asked God what was written in the contract. After hearing the details, each had to decline because of some difficulty in carrying out one or another of the commandments: the absolute bans on adultery, robbery, or killing, for example, could not be accepted, in good faith, as a binding commitment given the realities of life in this world. But at last God came to Israel,

> "And he (Moses) took the Book of the Covenant and read it to the people, and they said, whatever God spoke we will do and we will hear." (Exodus 24:7)

The right of Israel to the Torah was earned by unconditional acceptance: The other nations demanded to read the contract before signing; Israel signed before reading: "we will do now and hear (understand) later". If this be so, then it was Israel who did the choosing—not the chosen people, but the choosing people.

The Talmud in Shabbat, 88a, uses a metaphor of "seventy languages" to make the point that the Covenant of God was available to all nations and not only to Israel. The Talmud believed that there were a total of 70 languages spoken worldwide by humans. The Covenant was offered in all 70 tongues to all the world.

> Rabbi Yohanan said, What does scripture (Psalms 68:12) mean when it says: "God gave the word to a great host"? Every single word uttered by the Almighty was distributed in 70 tongues.
> It was taught in the study house of Rabbi Yishmael that the verse (Jeremiah 23:29) "as the hammer shatters the rock" means that just as a blow of a hammer strikes forth many sparks, so does every single word of the Holy One Blessed is He strikes out into 70 tongues.

So the Covenant was available not only to Israel alone, but only Israel accepted the Covenant uncritically. However, the Talmud is somewhat ambivalent about what may have been a hasty decision. Shabbat, 88a expresses the ambivalence through the words of a gentile critic.

> A gentile saw the sage Rava studying (Torah) with such passionate concentration that Rava was oblivious to the fact that he was digging his fingernails into his leg to the point of drawing blood. Said the gentile to Rava; You impetuous people: you put your mouths before your ears (you said "we will do" before "we will hear"), you yet persist in your impulsiveness. You should first have heard the details of the Covenant. If you found it convenient to carry it out—accept it; if not—reject it. Rava answered; we walk with integrity, unlike you people who seek advantage.

The story employs an interesting metaphor: intense study of the Torah can draw blood. Is the Talmud suggesting that the Torah could be viewed as a self-inflicted wound on the body of the Jewish people? What thinking person would ever enter into a contract without first hearing (understanding) the offer? Were there no Jewish lawyers at Sinai? Well, answers Rava, we are whole in our decision; which is not really a straight answer to the question.

Rava's exchange with the gentile can also be seen as a covert Christian-Jewish polemic. The word used in the story for *gentile* (*min*) can refer specifically to a Christian. The Christian points out to Rava that the Covenant between God and Israel (the Old Testament) has been superceded by a new Covenant (the New Testament); the commandments are no longer in effect, and it is only the impulsive stubbornness of Israel that persists. The essence of the connection to God is through the heart, and not in the performance of the old commandments. Rashi, the Talmudic commentator of the 11th Century, interprets the answer of Rava as a message of love understandable to both Jews and Christians; Rashi pins the Covenant to the heart:

> We walked with Him with a whole heart; in the way of lovers, we trusted Him that He would not burden us with more than we would be able to bear. (Rashi, Shabbat 88a).

According to Rashi, Rava answered that Israel accepted the Covenant without question because true love is marked by trust. But note that Rashi went beyond Rava in referring to the Covenant as a *burden*. Rashi, who witnessed the destruction of entire Jewish communities by the Crusaders on their way to purge Jerusalem of the Moslems, wonders aloud whether the burden of the Covenant is too much to bear. But Rashi, extending Rava, rejects the idea as unfitting for God. True love will never impose unbearable burdens. Our responsibility is to love God; His responsibility is to manage the world in a tolerable manner. Ruth too asked no questions when she clung to Naomi and accepted the God of

Israel [§52]. The question, however, remains without an answer: why did Israel, like Ruth, accept the Covenant unconditionally? What was the basis for such trust?

A most revealing interpretation of Israel's reason for accepting the yoke of the Covenant is also found in the deliberations of Shabbat, 88a:

> (Exodus 19:17, says) "And they stood at the foot of the mountain." (The expression "the foot of the mountain" is an English metaphor; the equivalent Hebrew metaphor is "under the mountain"). Rav Avdimi son of Hama son of Hasa said; (the expression "under the mountain" is to be taken literally) from this we learn that the Holy One Blessed is He lifted the mountain above them as one lifts a barrel and said to them; "If you accept the Torah—good, but if you don't, here will be your graves".

Now we know why Israel, when offered the Torah, agreed without reservation. Israel was no keener to receive the Torah than were the other nations; the difference was that Israel got an offer that could not be refused. Let me reconstruct Israel's response in colloquial terms: "OK, we'll do it, tell us the details later; just put down the mountain" (there were Jewish lawyers at Sinai after all).

We today might say that the people of Israel had no choice but to accept whatever God or Moses offered them at Sinai; what reasonable decisions could one expect from a ragged pack of people fifty days out of a slavery that had lasted hundreds of years? The picture of God holding a mountain over their heads might just be a metaphor describing the helplessness of the newly freed slaves. Metaphor or not, the Talmud draws a logical conclusion:

> Rav Aha son of Yaakov said; This situation provides an excuse for not living up to the Torah.

A forced contract is not binding.

Rava said; Nevertheless, they (Israel) willfully accepted it (the Covenant) later, during the days of Ahasuerus. As Scripture says (Esther 9:26), "The Jews accepted it and carried it out." They carried out that which they had already accepted.

Israel may have had no choice at Sinai, but Israel in her maturity did accept the Torah willingly, as recorded in the Book of Esther. Therefore, Israel can be held accountable to the Covenant. But we still are given no reason why Israel accepted the Covenant at that later time.

Indeed, there are serious problems dating Israel's binding commitment to the Covenant to "the days of Ahasuerus". The story of Esther took place about 800 years after the Exodus; did it take 800 years of turbulent Biblical history for Israel to take the Covenant to heart? On the contrary, the Bible itself records several occasions when Israel confirmed its commitment to the Covenant. Joshua renews the Covenant at least twice (Joshua 8:24-25; 24:18); Nehemiah reports in detail the ceremony at which those returning from the Babylonian exile accepted the Torah (Nehemiah chapters 8-10). One could cite additional occasions that could imply a confirmation of the Covenant. But most damaging to "the days of Ahasuerus" is that the citation from the Book of Esther has nothing to do with accepting the Torah; the Jews in Persia at the time merely committed themselves to observing the holiday of Purim. Purim celebrates the salvation of the exiled Jewish community in the Persian empire sometime in the fourth century BCE. Why did Rava ignore more plausible alternatives to "the days of Ahasuerus" for Israel's willful acceptance of its Covenant with God. Why, in fact, did Rava refer to Purim as "the days of Ahasuerus"? Ahasuerus is among the culprits of the story. The heroes of Purim are Mordecai and Esther (see the Book of Esther 1-10); why did

Rava not date the acceptance of the Covenant to "the days of Mordecai and Esther"? Rava is forcing us into interpretation.

Let me make a suggestion. The Book of Esther is the last book of the Jewish Bible to be admitted to the cannon. Esther marks the eclipse of God from the history of Israel; according to the Jewish tradition, after Esther prophecy was stilled, divine intervention became invisible, miracles were scarce, Heaven seemed quiescent. Israel was left to its own devices in confronting a hostile world. In fact, the Book of Esther is a story of the first political anti-Semitism: the Persian King Ahasuerus, at the bidding of his counselor Haman, has decided to wipe out the Jewish people.

> Their religion is different from all other nations, and they do not obey the laws of the King. (Esther 3:8).

Perhaps it was easier for Israel to keep the faith (more or less) for the first 800 years of Israel's history; God was sending His prophets and making His miracles. But "the days of Ahasuerus" marked the end of overt divine support. From then on, Israel had to leave a warm home and survive in a cold world. Since "the days of Ahasuerus", refuge from the threatening mountain could be had, not by accepting the Covenant, but by abandoning the Covenant. If Israel, despite the cost of being "different from all other nations", still persisted in clinging to the Covenant, says Rava, then Israel proved that she had accepted the Torah of her own free will. If you have willingly paid dearly for her/him/it, then you certainly must have wanted her/him/it.

In short, the Talmud teaches that the Torah was available, in one way or another, to every culture who might have cared to have it. Only Israel took up the challenge and held on, a hold that has lasted 3,000 years. More than God chose

Israel, Israel chose God. "The days of Ahasuerus", in the view of the Talmud, provided the true test and the proof of Israel's steadfast choice at Sinai.

We might also say that Rava's selection of "the days of Ahasuerus" for ratification of the Covenant speaks for the Talmud's realization of Judaism. Mordecai, the hero of "the days of Ahasuerus", is neither a prophet nor a priest; Mordecai is a sage. Mordecai persists in his unique Jewish identity, yet he overcomes the danger to him and his people by combining foresight, decisive action and faith. The Covenant is maintained by sagacious cognitive action, rather then by prophetic revelation or by priestly ritual. The religion of Israel has evolved from its days of Prophets in the desert and Priests in the Temple to become the Talmudic Judaism of human society. Rava appears in Shabbat, 88a, in a state of ecstasy; but unlike Ezekiel, the ecstasy of Rava is not in Heavenly Visions but rather in earthly study.

§54 Halakhah: Interpreting Torah

The binding code of Jewish behavior, as we said [§9, §10], is the *Halakhah*. Tradition has it that the *Halakhah* was received by Moses at Sinai as an *oral law*, as distinct from the Torah—Scripture, the *written law*. In fact, the Talmud as a whole is called *the oral law*. Thus, Shavuot marks the moment when Israel received the law in two distinct forms: written and oral. The Talmud in Menahot, 29b, tells it like this:

> Rav Yehuda said in the name of Rav; at the time Moses went up to heaven (from Mount Sinai to receive the Torah), he found God sitting and embellishing the letters of the Torah script with tags and crowns.

To this day, the font used to write the Torah scroll is highly ornamented; various of the letters are rich in tags and crowns, called serifs in the printing art.

> *Moses said before God (before God, and not to God; Moses is careful not to question God directly); Master of the Universe, why do you bother putting serifs on the letters?*
> *God said to him; one day in the distant future there will come a man, Akiva son of Yoseph is his name, who will use each serif to derive mountains upon mountains of Halakhah from the script of the Torah.*
> *Moses said before God; Master of the Universe, show him to me.*
> *God said to Moses; turn around.*

The scene shifts to the future Yeshiva of Rabbi Akiva. Moses enters, but

> Moses has to sit behind the eighth row because he cannot understand what they are talking about. Moses becomes distressed.

The bright students in the Yeshiva get to sit up front; the less talented sit in the back; Moses has to sit way in the back.

> At one point in the discussion in the Yeshiva, the students ask Rabbi Akiva; Rabbi, how do you know that your teaching is Halakhah (an authoritative ruling of the Law)?
> Rabbi Akiva said to the students; it is a Halakhah transmitted, generation to generation, from Moses at Sinai.
> Upon hearing this, Moses is relieved.

Moses is invoked by Akiva as the medium of the Divine code of behavior, the *Halakhah*. But hold on: if Moses is Akiva's authority, why is Moses such a poor student (a Yeshiva back-bencher); why can't Moses understand the Talmudic discussion that is attributed by Akiva to him, to Moses himself? No less puzzling: Why is the distress of poor-student Moses relieved when he hears that his revelation at Sinai is the source of the *Halakhah* teaching? How, in fact, does this scene answer the question put by Moses to God; why are You fiddling with serifs? I don't have a simple answer; but I imagine the Talmud is inviting us to think about *texts*, their *authorship*, *interpretation* and *truth*. A text can remain true to itself, despite the evolution of its meaning through renewed interpretation. Moses is still the source of Akiva's interpretation, even if Moses cannot understand Akiva's interpretation. The Torah can remain true only because it continues to live in interpretation. Moses is relieved because he sees that the Torah, through Akiva and his successors, will continue to live and grow in the vicissitudes of time and place [§78]. The Talmud is the fulfillment of the Covenant of the Torah; such would seem to be the lesson of this episode at Sinai.

§55 Ninth of Av: Day of loss

A few kilometers south of the Old City of Jerusalem is a high hill running East and West. The Haas Promenade is built along the crest of the hill, and on the north face of the hill below the crest winds the more pleasant Sherover Promenade. Both promenades provide a vantage point for contemplating the topography of Jerusalem and the Temple Mount. To the north, below the horizon, you can see the low hill of the Temple Mount with the golden Dome of the Rock, built in 690-1 CE by the Calif Abd-el-Malik, marking the site of the Jewish Temples, the First destroyed in 586 BCE and the Second destroyed in 70 CE. The Temple Mount lies before you within a natural amphitheater. The theater is enclosed on the east by the Mount of Olives and by Mount Scopus to the north; Scopus swings westward to join French Hill on the northern horizon. The west side of history's theater is formed by the hills of

the New City, including Abu Tor, the Hill of Evil Counsel. The theater is pierced to the east by the narrow Kidron Valley that meanders in a dry gorge through the Desert of Judea to the Dead Sea. If you were to fly like one of the black crows above the badlands of the desert, the Dead Sea would be only about 20 kilometers to the south-east and about 1,300 meters below the crest of the Haas Promenade, 400 meters below sea level.

Promenades are paths for progressing horizontally in space. The Sherover Promenade is wonderfully expressive; you can use it to look for a seat in this natural amphitheater. The Promenade begins at Abu Tor, half Arab village and half Israeli neighborhood. Walk south from there, pass the walls of the cloistered monastery of Saint Claire to your right, and traverse the olive trees and wild spice gardens on both sides of the path. You gradually climb towards the Haas Promenade on the hill to the south before you. To your left, you see the white badlands of the Judean Desert falling off to the Dead Sea and the heights of the trans-Jordan mountains of Moab rising beyond the great rift that houses the Jordan Valley and the Dead Sea. After about a kilometer, just as the Sherover Promenade turns east and prepares to end, you can look back and see the Dome of the Rock and the walls of the Old City. Here, resting on a bench you can interrupt your horizontal progress in space and begin a vertical promenade into time. With your body seated, your mind's eye can begin to wander over the terrain of time.

The Talmud telescoped the destruction of the Temple in 586 BCE together with the destruction of 70 CE into a single day of mourning: the 9th day of the Hebrew month of Av. The service is simple: a 24-hour fast with a chanting of the Book of Lamentations (Lamentations, chapter 1-4) added to the regular prayers.

The time is well chosen. The 9th of Av usually falls between the middle of July and the middle of August, when nature and patience in Israel are at their annual low points. There has been no rain for some months and none will come for some months. The landscape is dry, white-yellow-brown and bare; the heat is oppressive; the air is opaque, the sun has no mercy and the sky is bronze. The Hebrew word for summer is *kaitz*. The root, K-W/Y-TZ, denotes *the end* (*ketz*) and connotes the *end of life*. The word for thorn (*kotz*) derives from the same root; the supple green plant has become the sharp, dead thorn.

From your seat in the amphitheater, you can see David who took the city from the Jebusites in 1000 BCE, his son Solomon who built the First Temple there, the Babylonian assault that destroyed that Temple, and the Romans who destroyed the Second Temple 650 years later. You can see the site of the olive press (Gethsemane) on the Mount of Olives, the tower built by the Russian Tzar above it, the tower built by the German Kaiser to the north, and the tower of the Hebrew University farther north. But you can also see the others who have come to Jerusalem in vertical promenade: Egyptians, Greeks, Assyrians, Persians, Arabs, Crusaders, Mongols, and all the powers of Europe. The part of the world that has not come to Jerusalem has been molded by the part that has. Does the 9th day of Av mark the loss of Jerusalem or the gain of Jerusalem?

§56 Rosh Hashanah: Day of judgment

The Hebrew month of the autumnal equinox is Tishre, which usually falls out in the second half of September. A subtle change in the atmosphere of Israel heralds the approach of Tishre: the sky at noon is less metallic, the wind freshens the late afternoon, the clouds pile up in the west and the sunset lights them in rose, the earth emits a fresher smell; one may see a wagtail freshly migrated from Europe, and tall stalks of autumn flowers appear on the way to Jerusalem. The season is changing; the rains are on the way.

Leviticus 23:23-24 says enigmatically;

> And YHWH spoke to Moses saying: Speak to the Children of Israel saying; on the first day of the seventh month you will make a Sabbath day, a memory of trumpeting, a holy convocation.

The Talmud interprets this holy convocation to be Rosh Hashanah: the *head* (*rosh*) of the *year* (*shanah*; from the root SH-N-H, which we learned above means to *repeat*, to *study* and to *cycle* [§9]). Thus, the first day of the seventh month (Tishre), according to the Talmud, marks the Jewish New Year—the anniversary of the Creation of the World, the Creation of Adam and Eve, and the Day of Judgment. Rosh Hashanah, which the Talmud extended to two days, is the first of the two High Holidays, or Days of Awe; the second Day of Awe is Yom Kippur, the Day of Atonement, which comes ten days after Rosh Hashanah.

The Mishnah in tractate Rosh Hashanah, 16a (the name of the holiday is also the name of a volume of the Talmud), puts it this way:

> The world is judged in four chapters: the grain on Passover; the fruit of the trees on Shavuot; on Rosh Hashanah, all who walk the earth pass before Him as benei maron (an obscure phrase that will be clarified below), as it is written (Psalms 33:15); He Who created their hearts alike, He examines all their deeds; and the rain on Sukkot.

What does it mean, "the world is judged"? How can nature be judged? The Talmud goes on to explain that of all the world, only humankind

JUDGMENT

is judged; only humankind is empowered to choose between right and wrong. Only choice needs judgment, so human action alone stands judgment. Thus, humans are judged, and that judgment determines the success or failure to satisfy human needs of the grain, the fruit and the rain.

Here is a revolution in worldview. The religions of natural time, as we discussed above, understood that nature, personified as fate and the whims of the gods, determines human achievement [§34, §35, §36]. Humans, like all things natural, are trapped in nature. Judaism turns the relationship between humankind and nature on its head. Humankind is not ruled by nature; on the contrary, humankind determines the course of nature. The rain and the crops fail or succeed according to God's judgment of women and men. We discussed above the West's idea that humans have the right to intervene in nature [§2]; Rosh Hashanah, extending the idea, teaches that humans, by their actions, are *responsible* for nature. What might seem to humans to be the chaotic behavior of world is a just punishment for human vice. The right behavior of humans leads to the right behavior of nature. A rational world *requires* a rational humanity.

The idea that mortal men and women could be responsible for nature probably seemed ridiculous to most people 2000 years ago. Today, we see that humans do indeed determine the course of nature: global warming, pollution, the disappearance of the rain forest, the extinction of creatures, the spreading wasteland. The idea that nature was dependant on human morality was not unique to Judaism, but Judaism endowed the individual human with special responsibility for the state of the world. Rosh Hashanah annually renews this responsibility. If the universe has been created for you and

for me [§14], then what you or I choose to do creates a world. If we do good, we have made a good world; if we do bad, we have created a bad world. The act of judgment is the act of looking at the worlds each of us creates: the good and the bad.

The Talmud develops further the idea of responsibility and judgment in Rosh Hashanah, 18a. The discussion begins by exploring the possible meanings of the obscure words "benei maron" that appear in the Mishnah:

> "On Rosh Hashanah all who walk the earth pass before Him as benei maron"; what does "benei maron" mean?
> Here (in Babylonia) we translate the term "benei maron" as imarna "sheep".

The judgment of humanity on Rosh Hashanah is likened to a file of sheep passing one-by-one through the turnstile as the Shepherd judges the worth of each. Some sheep are selected by the good and knowing Shepherd for continued care and others are selected for slaughter.

> Resh Lakish said; "benei maron" refers to the steep path that leads to the village of Maron.

Maron was a mountain village in the upper Galilee that could be reached only by traversing a narrow path bound on either side by a deep gorge. The picture of judgment proposed by Resh Lakish suggests that God judges how each of us navigates the narrow ridge of right living without stumbling off the path into disaster. Resh Lakish had been a gladiator and bandit before he became a Rabbi (Baba Metzia, 74a), and must have been familiar with living a step from death.

> Rav Yehuda in the name of Shmuel said; the term "benei maron" refers to the soldiers of the House of David.

The Steinsaltz Talmud (cited above; §8) comments in Rosh Hashanah, 18a, that the problematic term *"benei maron"* may be a Hebrew-Latin corruption of *"be-numeron"*, the mustering of troops *by their numbers*. The military metaphor of Shmuel thus has an etymological basis. The connotation is that one's life stands muster through one's performance as a warrior in the cause of God.

The three Rabbis use an obscure set of words in a Mishnah text "benei maron", possibly a corrupted spelling, as a vehicle for exploring the meaning of Rosh Hashanah. Common to the three different situations is the picture of a moving file of subjects observed by the eye of the Shepherd, the Commander, or the distant Witness. The picture of a file invites us to consider the individual within the collective. The Talmud continues:

> Rabba son of Bar Hana said in the name of Rabbi Yohanan; the entire file is examined in one glance.

The file of those who "walk the earth", whether as sheep, as walkers at the edge of doom, or as warriors (or visiting scientists [§1]), are seen by the Judge as a group. Responsibility is social as well as individual.

> Rav Nahman son of Yitzhak said; we too have learned an interpretation of the verse (Psalms 33:15); He Who created their hearts alike, He examines all their deeds. What does the verse mean? If it intends to say that the Creator fashioned all human hearts the same, that clearly is not the case.

Humans are individuals and each human heart is individual [§1].

> The verse intends to say; The Creator sees the heart of each individual within the collective at once, and He examines all their deeds.

Each person stands in judgment as an individual *within* a collective: How do *I* look to the Judge? How do *we* look to the Judge? As we file by, the two questions are one. Each of us is an individual responsible for his or her society.

§57 Torah texts for Rosh Hashanah

The Biblical texts selected for chanting during the prayer service on Rosh Hashanah fashion a meta-textual commentary on the meaning of the holiday. Rosh Hashanah is the birthday of the universe, but we do not read the creation saga of Genesis that opens the Torah. No, we read stories about the birth of only children and about man and woman, the unspoken relationship between the couple and the life of their only child. The creation of the universe on Rosh Hashanah is the birth and salvation of a single child within a family [§14]. Below, I list the selections for the two days of Rosh Hashanah; read them and interpret their message.

On day one we read Genesis 21: 1-34: the story of the birth of Isaac to the barren Sarah; the conflict between Sarah and Hagar that leads to the ejection of Hagar and her son Ishmael into the desert; and the salvation of mother and son through the prayer of the child. The selection from the Prophets is First Samuel 1: 1-28, 2: 1-10: the story of the birth of Samuel to the barren Hannah and Hannah's song of praise to God.

Both stories are the stories of women. In each story, the husband of the initially barren woman has a second wife who has given birth. Abraham, Sarah's husband, has had his child Ishmael through his second wife Hagar, and Elkanah, Hannah's husband, has had many children through his second wife Pnina. In each story, the man's love is greater for his barren wife. Jealousy, rivalry, frustration, hate, love, failure and triumph seethe below the surface, as they always do in the Biblical narrative. Each of the only children born to Sarah and Hannah are destined to leave home for a "greater good".

On day two we read Genesis 22: 1-24: the story of God's command to Abraham to sacrifice his beloved son, Sarah's only child, Isaac. We witness Abraham's silent acquiescence, and we are relieved by the sacrifice in Isaac's stead of a ram caught in the thicket. This passage is exalted; it is no wonder that Erich Auerbach chose it to exemplify the roots of the West's perception of reality, in contrast to Homer [§3].

But the Talmud does not accept the happy end. The Talmud goes beyond the Biblical story and tells us about the anguish of the mother: Satan rushes to Sarah in the guise of her son Isaac, and Isaac/Satan recounts his close call with death. Sarah, upon hearing her son's words, falls dead (Midrash Tanhuma, cited in *Sefer Ha'Aggadah*, by Haim N. Bialik and Y. H. Ravnitzky, Devir, Tel Aviv, 1956, page 32). Abraham, covering for God, had hid from his wife Sarah God's commandment to sacrifice their son Isaac. But the Talmud never covers up for God; the end is never without backlash. Sarah is dead before Abraham can make it home to explain how it all worked out for the best.

The selection from the Prophets on day two is Jeremiah 31: 1-19: The prophesy that Israel, after the destruction of the Temple and Exile to Babylon, will be remembered as a beloved child by YHWH. Israel, like beloved son Isaac, will be redeemed: parents and children, a recurring theme in Judaism.

§58 Shofar: The trumpet

Leviticus 23:23-24, you recall, refers to the first day of the month of Tishre as a holiday of *memory* and *trumpeting* [§56]. The Talmud integrates the themes of *memories* and *trumpets*, along with the *kingship* of YHWH, into the Rosh Hashanah prayers.

Rabbi Yehuda said in the name of Rabbi Akiva; ....and say before Me on Rosh Hashanah (in the prayers) verses of kingship, memories and trumpets: kingship—so that you will make Me your King; memories—so that your memories will come before Me for your good; and by what means? By the trumpet (Rosh Hashanah, 16a).

Kingship, *memories* and *trumpeting* are the terms for three collections of verses said during one of the major prayers of the holiday. Each of the three subjects is represented by 10 Biblical verses—three verses from the Torah, three from the Psalms, three from the Prophets, and a final verse again from the Torah.

The *trumpeting*, which is the third of the three themes, is represented not only by Biblical verses; trumpeting is expressed concretely by the blowing of the ram's horn, the *shofar*. The *shofar* is a simple instrument originating in human pre-history; a hollow ram's horn that can be made without much fuss. Extracting a decent sound from the horn, however, does demand considerable expertise. The *shofar* is blown with great ceremony at appointed times during the prayers on Rosh Hashanah. Each person is commanded to hear the sound of the *shofar* (again, Judaism values *hearing*, rather than *seeing* [§9]). The act of blowing and the act of hearing both require intention and attention (Rosh Hashanah, 26ab-27ab). What does hearing the *shofar* blast mean?

Actually, the *shofar*, other than being an emphatic sign of Rosh Hashanah, is a signifier that is not clearly anchored to a specific signification. Like the symbols of Passover [§46], the *shofar* can mean different things to different people. The sound of the *shofar* is an impressive reminder; but it can remind us of variously different things at once: the ram that was sacrificed by Abraham in place of his son Isaac; the sound of the trumpet blast when the Torah was given at Sinai; the call to battle; the Jubilee—freedom from slavery; the final redemption. Perhaps most vivid is the child's memory of a grandmother in her outmoded finery hurrying in awe to the synagogue to hear the blowing of the *shofar*, thinking of her childhood memory of her own grandmother hurrying in awe to the synagogue.

§59 Memories

Memory is the second theme proposed by Rabbi Akiva for the Rosh Hashanah prayers. The concept of memory fits the day; what is the individual if not a unique memory? Each of us is unique in his and her memories, which are the imprint on each mind of one's unique experiences in life [§1]. And each of us strives to leave a memory of our unique presence in the world. Memories are both interior and exterior—what each, on the inside, remembers of his or her world and what the world, on the outside, remembers of us.

So what does Rabbi Akiva mean when he teaches that verses of memory are to be said "so that your memories will come before God for the good"? The memory verses quoted in the prayer book refer to God's remembering of His Covenant with us. But who are we to remind God to remember? God is the One Who never forgets. Perhaps the Talmud would like us to remember just who we are. What mark have we made, both inside and outside ourselves? Judgment is not of acts, but of memories of acts and their consequences. Memory is a beginning of time.

§60 Kingship

The verses of *kingship* recited on Rosh Hashanah, the third theme of Akiva, refer to YHWH as King of the Universe. Why does the Talmud

bother to assign a special holiday to the assertion of God's kingship? After all, reciting the *Shma* twice a day declares that YHWH is our God (Elohim) [§39]. If we state that YHWH is God, is that not the same as stating that YHWH is King of the Universe? God and King of the Universe would seem to be redundant terms; or are they? The Talmud records a difference of opinion on the matter (Rosh Hashanah, 32b): Rabbi Yossi says there is no difference between the two declarations, and Rabbi Yehuda says there is a difference. Neither of them explains his point of view. How might we interpret the difference between YHWH as God (in the *Shma*) and YHWH as King of the Universe (on Rosh Hashanah)?

The difference arises from the different ways the world affects our sensibilities: we see both the order of the world and its chaos. The name God (Elohim) refers to YHWH as the source of the law and the order. The immutable laws of physics exemplify the order of the material world. Justice exemplifies order in the realm of human affairs. Right is right and wrong is wrong; such is the just order of events. From this point of view, Elohim (God) is a job description. YHWH as Elohim establishes material order and social justice. Justice, like a fundamental law of nature, is a fact subject to proof. Justice and laws of nature are always moot: you can test a law of nature by experiment; you can plead for justice before a court. Remember how Abraham challenges God's decision to destroy Sodom and Gomorrah (Genesis 18:25); "Far be it from You to kill the righteous with the wicked.....Will the Judge of all the world not do justice"? Justice, like experimentation, is subject to public examination. Justice has standards. YHWH as Elohim is accountable both for the laws of nature and for the just management of the world.

The King of the Universe, in contrast to Elohim, relates to the unjust aspects of existence. *Accountability* expresses the difference between YHWH as Elohim (God) and YHWH as King of the Universe. A king is really a king when he is accountable to none but his own will. The only kings remaining today are constitutional monarchs or mere figureheads; a true king is an absolute ruler and not subject to public examination. Thus, the King of the Universe is Absolute Power. The King of the Universe is the aspect of YHWH beyond our understanding, beyond our sense of justice. The King of the Universe is the Author of what we perceive to be the ultimate mystery of life and death.

Rosh Hashanah marks the creation of the world, the birth of the individual and the judgment of each. But not a single one of us is responsible for the time, the place, or the circumstances of his or her birth. These facts of life, however fortunate or unfortunate, are not dictated by justice. Our beginnings and our ends are the unaccountable choice of the King. Life and judgment are thrust upon us; so too is death. Is there Justice in the human condition? The individual and the collective, in crowning YHWH as King on Rosh Hashanah, accept the terms of human existence. Rosh Hashanah celebrates the beginning of time with our recognition of the human condition, and with our acceptance of its irrationalities.

Above, we discussed how the natural religion of mythology explained the irrational and unjust by the whims of the gods and by fate—what we called the two faces of nature [§35]. Monotheism, in contrast to natural paganism, insists that the One God is supremely rational and without internal division. But how then can Judaism explain the tragedies of existence? Rosh Hashanah does not explain them; the day teaches us to accept them.

Rosh Hashanah helps us interpret the formula of the *brakha*, the acknowledgement of time discussed above [§44]:

> Acknowledged are you YHWH, our God (Elohim), King of the universe who has commanded...

The *brakha* mentions three aspects of the One: YHWH, God, King.

YHWH is a personal name and so expresses the personal bond of each conscious being with the Being of the Universe—YHWH is one's personal agency of mercy and compassion, one's Father. YHWH is addressed as *You.*

God (Elohim) is the lawful and just manager of creation. Elohim is addressed as *Our God*, to whom we turn for justice.

The King is Absolute. We address the King in the distant third person.

We can speak to YHWH. We can understand Elohim. The King is beyond understanding.

Although the *brakha* contains three terms, a major prayer of Rosh Hashanah refers only to two of the terms: "Our Father, Our King". The orderly and just progression of the world is automatic and needs no prayer. Our personal intervention, prayer, is relevant to the personal aspect of YHWH (Our Father) and to the aspect of YHWH beyond understanding (Our King).

§61 Yom Kippur: Day of atonement

The judgment that begins on Rosh Hashanah is sealed 10 days later on Yom Kippur, the Day of Atonement. The Mishnah in Yoma, 85b states:

> Yom Kippur atones for transgressions between man and God. Yom Kippur cannot atone for transgressions between a man and his fellow, until the transgressor propitiates his fellow. (Man and his fellow in Hebrew include women.)

ATONEMENT

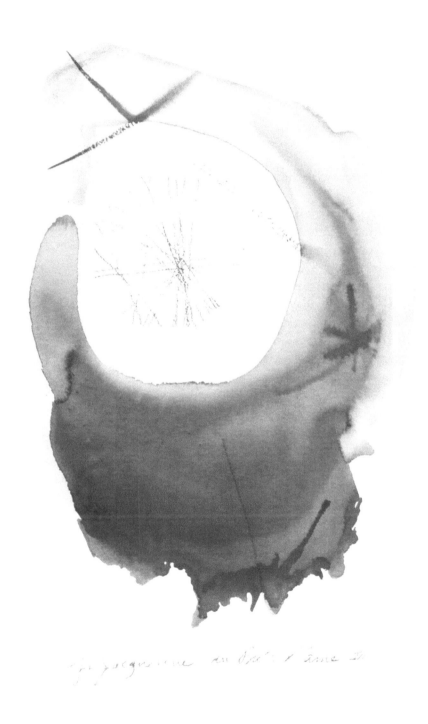

What does the Talmud mean when it says that "Yom Kippur atones for transgressions..."? Atonement is done by people; how can a day atone? To answer the question, let us analyze the process we call atonement. Individual atonement would seem to require three elements:

- Regret for the wrong;
- Abstention from repeating the wrong;
- Restitution for the consequences of the wrong.

How does the Talmud deal with these three elements?

Regret is manifested by public confession, which on Yom Kippur is recited by the congregation in unison on five different occasions during the 24-hour day of prayer (see Yoma, 87b). The Talmud discusses various formulae of public confession, but the standard form used today is composed of an alphabetical list of faults, two faults for each of the 22 letters of the Hebrew alphabet.

The alphabetical format in Hebrew is ontological. The letters of the alphabet, by recombination, form words. And words are the coin of human cognition. The alphabet, therefore, represents the primal sounds from which humans construct the words that create understanding. The alphabet encodes the portion of reality accessible to humans—remember, davar is both a spoken word and a thing [§9]. So any alphabetical arrangement of a Hebrew text implies a combinatorial universe beyond the given combination of the actual words used in the particular text. An alphabetical text is all-inclusive, potentially. Thus, an alphabetical confession includes the universe of sins, beyond those actually mentioned, that might be constructed through mixing various combinations of the various letters. Each person regrets his or her personal list of sins through recitation of the public alphabetical list; mix and match.

The content of the confessional list is interesting. Of the 44 listed faults, at least 11 can be related to improper verbal expression: unkind words, wrong speech, scorn, impure speech, foolish words, denial, crude joking, slander, unseemly conversation, gossip, false testimony. Language is the uniquely human attribute, and the Yom Kippur confessional sharpens our sensitivity to how we misuse our human mark.

The other faults to be confessed are also universally human, and not unique to the *Halakhah* [§9, §10]: willful misbehavior, hardheartedness, lack of knowledge (we are responsible for using our brains effectively; how Talmudic), wanton sex, wrongs done in public and in private, cheating, misleading our fellow, fantasies (cognitive misuse), prostitution, disrespect for parents and teachers, wrongs committed by mistake as well as by premeditation (the Rabbis knew about the Freudian slip), aggressive behavior, profaning the holy, wrongs done through misunderstanding (another misuse of the brain), bribery, misleading negotiations, unhealthy intake of food and drink, usury, pride, unruly eyes, self-importance, stubbornness, hate, naïveté (more cognitive misuse), and so on. One is not asked to confess his or her failure to be a saint; one is obliged to confess on Yom Kippur one's failure to be a substantial human being—a *Mensch* (Yiddish and American English). Indeed, among the many Hebrew words for *sin*, the word used in the formula of the Yom Kippur confessional is *het*, a word that means to *miss-the-target*. We regret being off the mark to which we humans all aim.

Abstention in the face of opportunity, the second element of atonement, is defined by the Talmud thusly:

> How does one demonstrate repentance? Rav Yehuda said; if the person confronts the sinful situation once and twice more and does not succumb. Rav Yehuda

provides an example of abstinence despite meeting the same woman, at the same time of night, and in the same place (Yoma, 86b).

Restitution, the third element of atonement, is clear with regard to a transgression against a fellow person: payment for the material damages caused to the person and conciliation for his or her shame, the damages to self-esteem (Yoma, 87a).

§62 Yom Kippur: Day beyond time

Regret, abstention and *restitution* are performed by the atoning person; but the Talmud, as we pointed out above [§61], states that the day itself atones. What element can the day supply to the process of atonement? How can time effect atonement? Consider that the day, time itself, is required for *restitution*. Look at the problem this way: Above, we saw that *existential time* is created by the acts we will to perform (§38). Individual acts form individual time. Goodness makes good time. A wrong act, in contrast, warps time. Our faults damage time itself. One makes restitution to a hurt fellow by payment and conciliation; the restitution of damaged time is the cosmic role of the day of Yom Kippur.

The volume of the Talmud dealing with Yom Kippur is named *Yoma*, an Aramaic word that simply means *The Day*. Yom Kippur is The Day distinct from all other days because it is the day *outside* of existential time. Obviously, chronological and natural time do proceed on Yom Kippur. The world goes on. But existential time stops for those who observe The Day. Yom Kippur is characterized by the cessation of all the creatural activities that maintain human life and allow the creation of existential time: no eating, no drinking, no bathing, no anointing of the body, no sexual union, no shoes. The needs of bodily metabolism, personal adornment,

procreation, and movement through the world are suspended for a full day—from before sunset to after sunset of the next day. The individual and the community leave the affairs of this world to observe Yom Kippur.

Yom Kippur in Israel is noticeable by the sudden silencing of the din of radio, TV, street traffic, air traffic; the country shuts off the noise of modernity. You begin to hear the wind in the trees and the bird-song. Free of food and drink, your senses sharpen and you begin to smell the scent of the earth. The devotion of a day of one's existential time to contemplation of atonement is one's restitution to time wasted on misdeeds. Yom Kippur calls a halt in creatural time to repair existential time. So I have come to see it; other interpretations are welcome.

The story of the prophet Jonah (Jonah 1-4) is read on Yom Kippur in the afternoon prayer. The afternoon prayer is relatively short and low-key compared to the emotional heights reached by the morning and final prayers. The book of Jonah, to my mind, is the focal point of the afternoon prayer, and Jonah sets the stage for the closing prayer that follows. Jonah is a cinematic portrayal of the complexities of atonement, both for individuals and for communities.

Jonah is Everyman: a prophet who hears the word of God and yet remains a petty, self-centered, misanthrope who lacks the vision to see the wonder of his own life. The story opens with God commanding Jonah, a Hebrew, to warn the people of Nineveh to mend their evil ways. Nineveh is a great metropolis to the East inhabited by pagans. The story is about a Jew with a message for the gentile world. (Is Israel thus questioning its universal mission personified as the man Jonah?)

(The word *jonah*, by the way, means *dove*, and so Jonah through his name connotes the dove with the olive sprig who signaled peace to Noah in the Ark; the earth was now ready to receive Humanity after the Flood; see Genesis 8:8-12. The dove signifies love in the Song of Songs; 1:15; 2:14; 4:1; 5:2, 12; 6:9.)

Jonah tries to flee his calling by sailing in a gentile ship to distant Tarshish (Tarshish is probably near Gibraltar, at the far end of the known world). We are not told why Jonah wants out; is he afraid of ridicule or anti-Semitism, or is he merely a disbeliever? God sends a great storm that threatens to sink the boat, and Jonah is found out by his shipmates. To save the lives of the crew, Jonah is cast into the raging sea [§21]. The sailors are astonished to see the sea calmed and make atonement for having cast out Jonah; the sailors make their private Yom Kippur at sea. Meanwhile, Jonah makes his atonement, spending his own three-day Yom Kippur in the belly of the great fish prepared by God. Saved from the fish, Jonah returns to his duty, journeys to Nineveh and warns the people to repent. To Jonah's astonishment, the people of Nineveh respond to the call and make atonement through prayer and public fasting—a kind of public Yom Kippur. God forgives the people of Nineveh, and decides not to destroy the great city.

Instead of rejoicing at the good he has wrought, Jonah is bitter because he has lost face. The embittered Jonah retires from the city to sit under a shelter (*sukkah*) while waiting to see what will happen (Jonah 4:5). God sends a desert wind from the east to punish Jonah, and then saves him from the burning sun with the shade of a miraculous gourd tree. Then God turns the wheel of fate and sends a worm to kill Jonah's gourd tree. But Jonah, in contrast to the people of Nineveh, fails to understand God's message. And so then God speaks openly to his poor prophet Jonah sitting there in his flimsy *sukkah*;

you Jonah show more mercy for the shade of the gourd tree than you show for the myriads of people and animals that inhabit Nineveh.

How can the dense, unfeeling and unknowing Jonah serve as the instrument to save a boat-full of sailors and a great city of many hundreds of thousands of human beings? Despite himself, he can and he does. Jonah is not among the most inspiring of the prophets of Israel; he is no Samuel, Amos, Jeremiah or Isaiah. But as weak as he is, is Jonah not the most human? Is there a better Talmudic commentary on Yom Kippur than is our Jonah? How many of us, like simple Jonah, sit in our flimsy shelters—*sukkot*—waiting for the gourd tree, the worm and destruction or salvation?

At the close of Yom Kippur, the Torah (as interpreted by the Talmud) commands each person to build a *sukkah* to celebrate the holiday of Sukkot, which comes five days after Yom Kippur.

§63 Sukkot: Returning to time

Traditionally, a Jew begins building the *sukkah* for the Sukkot holiday immediately upon returning home from the synagogue at the end of Yom Kippur. A *sukkah*, when it is not used as a ritual on Sukkot, is a shelter that until recently could be seen each autumn in the orchards and vineyards of Israel when the grapes, dates, and pomegranates are harvested. The word *sukkah* simply means a *shelter*, a harvest shelter. The *sukkah*, in nature, signifies the harvest season because it serves as a temporary watchtower, storehouse and resting place during the harvest. The holiday of Sukkot originated as a local version of the universal holiday of the harvest, an Oktoberfest; see *The Golden Bough. A Study in Magic and Religion*, by J. G. Frazer, cited above [§22].

Sukkot, like all harvest festivals, is a holiday of joy. The dry summer is over, the wine is good, and the rains are on the way to revive nature. The palm frond, the citron fruit, the willow branch and the sprig of myrtle are taken in hand and blessed as signifiers of the harvest and the coming rains (Leviticus 23:40).

But the Torah, in its creation of historical time, transforms the *sukkah* from a natural signifier of the harvest into an historical signifier of the Exodus (Leviticus 23: 42-43). Dwell for seven days in *sukkot,*

> So that your generations will know that I made Israel dwell in shelters (sukkot) when I took them out of the land of Egypt; I am YHWH, your God.

The nature holiday of Sukkot becomes a link in the chain of Israel's history. The Talmud fashions the mundane harvest shelter into an existential ritual by prescribing in great detail the *sukkah's* size, material construction, situation in the environment and human occupation during the holiday of Sukkot. An entire volume of the Talmud is devoted to the *sukkah* (Tractate Sukkah; see *Old Wine, New Flasks: Reflections on Science and Jewish Tradition*, by Roald Hoffmann and Shira Leibowitz Schmidt, W. H. Freeman, New York, 1997, pp 62-68). Humans, deprived of feathers or fur, need shelter for survival; the *sukkah* is a signifier of the precarious human condition and also of the human's ability to solve material problems using technology.

The ritual of the *sukkah,* through its connection to Yom Kippur, goes beyond historical time and becomes a metaphor of existential time. Yom Kippur, as we discussed, suspends time; Sukkot marks the return from atonement to worldly time: build a house (the *sukkah*), harvest the grapes, drink the wine, eat the dates, bless the greens, prepare for the rains and renew the cycle of life. The *sukkah,* the shelter we build on our return from Yom Kippur to this world, is a signifier of human ingenuity.

But the *sukkah,* like one's body, is a fragile and transitory shelter. The *Halakhah* [§10] requires that the roof admit both the rain and the sunlight, both the cold of night and the heat of day. You sit in your *sukkah* as the prophet Jonah sat in his *sukkah* (Jonah 4:5), exposed to the elements and waiting to see what God will do to you and to the people of Nineveh.

(Avodah Zara, 3b, tells that the Nations—Romans, Persians and others—complained to God they too would have accepted the Covenant had He forced it upon them with Mount Sinai over their heads as He had forced the Covenant upon Israel. God, out of fairness, agrees to give the Nations one "easy" commandment to fulfill: build a *sukkah* and celebrate *Sukkot.* The Nations, however, abandon the *sukkah* because they can't take the heat in such a flimsy shelter. Israel, in contrast to the Nations, fulfills the Covenant by sitting in the *sukkah* and taking the heat [§53].)

Sukkot terminates the Torah's cycle of natural time made historical: Passover [§47], Shavuot [§51], and Sukkot [§63]. Cycles, by their nature, go on and on; Sukkot also initiates the beginning of the new cycle. The duration of Sukkot is meaningful. Sukkot is basically a holiday of seven days (Numbers 29: 12) with an additional *eighth* day appended (Numbers 29: 35). Eight supercedes seven. The eighth day breaks the seven-day cycles—the Sabbath week, the seven days of Passover, the seven times seven days between Passover and Shavuot. The eighth day looks ahead to the new cycle; the eighth day is the beginning of a new week and of a new world. The eighth day is the day after the six-day creation of the world and the Sabbath day of rest. The prayer that blesses the return of the rains of life to the dry Middle East are said on the eighth day of Sukkot.

And the eighth day of Sukkot is the day on which the synagogue service ends the yearly cycle of the reading of the Torah and begins the cycle anew. Every Sabbath service includes the ceremonial reading of a consecutive portion of the Torah; the 52 weeks of the year cover the Five Books of Moses from beginning to end, commencing and terminating on the eighth day of Sukkot. Sukkot is the rebirth of worldly life and time.

The 9th of Av, the day of the destruction of the Temple, which we discussed above [§55], is a memory of history that did not originate in a holiday of nature. Hanukkah and Purim are two additional holidays outside of the natural cycle.

§64 Hanukkah: Days of light

Judaism, as we have discussed, is unique in transforming nature into history [§37]. Except for Hanukkah. Hanukkah is the one holiday in which the Talmud transposes a truly historical event into a mythic history, and then proceeds to connect the history-based ritual to a cycle of nature. Hanukkah is counterpoint to the Judaic historization of nature. Hanukkah is the exception that illustrates the rule by breaking it.

Hanukkah is rooted in the history of a series of military victories led by Judah the Maccabee over Greek armies. The campaign culminated in 165 BCE. Judah's victory made it possible for Judaism to resist assimilation into the cultural melting pot of Hellenism that unified the Eastern Mediterranean world in the wake of the conquests of Alexander the Great. The Maccabean victory made it possible for the Jews to preserve their cultural uniqueness, and so the Maccabean victory made possible the later emergence of Christianity and Islam. Hanukkah signifies an event that *made* history. How does the Talmud interpret this history?

Surprisingly, the Talmud pays little attention to the military victory. Instead, the Talmud emphasizes the purging of the pagan Greek ritual from the Temple in Jerusalem and the restoration of the traditional Jewish ritual. The story of Hanukkah is recounted in two versions of the Book of the Maccabees (Maccabees 1 and 2). The Rabbis of the Talmud, however, did not accept either book of Maccabees into the Biblical Cannon, and Maccabees 1 and 2 have survived only in Greek translation in the Apocrypha.

The word *Hanukkah* is usually translated as *dedication*, the dedication of the Temple. But the English word *dedication* has a different connotation than the Hebrew *Hanukkah*. To dedicate is *to proclaim, to set aside, to devote*; to dedicate the Temple in English is to make the Temple holy. The root of Hanukkah is H-N-KH, which means *to teach, to educate, to construct, to work to add value beyond the natural endowment*; to dedicate the Temple in Hebrew is to work to establish its unique character.

The ritual of Hanukkah plays down the historical moment in favor of a miracle involving a little jug of olive oil. After the military victory over the invading armies, it seems that the restoration of the Jewish Temple ritual was in danger of being delayed for lack of the ritually pure olive oil needed to light the candelabrum (*menorah*; see Leviticus 24:1-4). Miraculously, a small jug of kosher olive oil was found, but, alas, the amount was sufficient to light the *menorah* for no more than one day. Nevertheless, that meager amount of oil sufficed to fuel the *menorah* for 8 days. The 8 days provided sufficient time to prepare new olive oil to maintain the holy light. To this day, Jews celebrate Hanukkah by lighting oil lamps (or candles) for 8 days. Hanukkah is the Holiday of Light.

(Is it only a coincidence that Hanukkah, like Sukkot, lasts for *eight* days? Re-dedication is the beginning of a new cycle.)

But why did the Rabbis play up the jug of oil and play down the miraculous victory of Judah the Maccabee? Without Judah there would have been no Judaism. As we saw at the disputation about Akhnai's oven [§21], the Talmud was not greatly impressed by miracles; it seems foreign to the spirit of the Talmud to focus on the jug of oil. We would have to conclude that the Talmud is even less impressed by military power than it is by miracles. The conquest of Canaan is not celebrated by a Jewish holiday, in contrast to the Exodus of the slaves (Passover), the receiving of the Torah (Shavuot) and the wandering in the desert (Sukkot). Clearly, the Rabbis were not enthralled by the High Priest kings of the Hasmonean dynasty that arose from the Maccabees; the Hasmonean kings of Israel struggled for power like their despotic neighbors, sought the favor of Rome, and were finally ousted by the House of Herod. The Talmud preferred the miracle of the little jug of oil to the military history of the Maccabean revolt and its aftermath. Mythic history is always more attractive than are current events. Talmudic Judaism emerged from the ashes of the Second Temple [§6]. The actual history of Hanukkah, the victory that restored the Temple ritual, did not suit the spirit of the Talmud. The Rabbis who created the Talmud, in fact, replaced the Temple ritual and its priesthood with the spirit of Judaism described in this book.

The Holiday of Light metaphor refers both to the miracle of the jug of oil and to the triumph of spiritual light over dark superstition. But the Hanukkah light also connects history to nature. Hanukkah is observed at the time of the winter solstice—the day in mid-winter when the sun ceases its daily decline and again begins to rise in the dome of the sky; the hours of light (the day) begin to re-conquer the hours of darkness (the night). Many peoples light fires at Hanukkah-time-of-year, not just the Jews. The Yule log of Christmas too marks the universal holiday of light (see *The Golden Bough. A Study in Magic and Religion*, by J. G. Frazer, cited above [§22]).

§65 Purim: Coming of age

Purim, as we discussed above [§53], suits the Talmudic worldview very well. Unlike Hanukkah, Purim has no military conquests, no Temple or priesthood, and no overt miracles. Purim is making do in Exile. The Book of Esther was accepted into the Biblical Canon, while the heroic Book of the Maccabees was not.

§66 Does the world make sense?

I set out to connect ideas common to Western science and to Judaism using Talmudic texts that probe the value of the individual in nature (Person), that empower humans to intervene in nature (Place) and that command human progress (Time) [§2]. Science emerges from the axiomatic belief that the world makes sense—that events have reasonable explanations. The rationality of the world is a corollary to the idea of One God [§39]. Monotheism, as developed by Judaism, believes that One Individual, YHWH, is in charge. The world is a text written by an Author. The narrative of nature is the unfolding of history. The monotheist trusts the Author of history to be reasonable. Polytheism, as we have discussed [§35], does not endow nature with reason; fate and chaotic powers rule human life, personified by the gods and their fateful whims. A multi-authored narrative has many conflicting

SUFFERING

aims. Polytheism, in contrast to monotheism, is unruffled by the injustice of existence.

But how does Talmudic Judaism explain suffering? Granted that YHWH is the unfathomable King of the Universe [§60], how is the individual to respond in practice to what seems to be unjust? Even though we cannot presume to know the mind of God, we need to satisfy the quest of our own minds to understand the misery we see or feel. We shall end this section with a Talmudic text that explores undeserved suffering—the rub of monotheism. As you might expect, this problem is addressed where the issue is most acute—in the context of the *Shma*, the declaration that YHWH is the Singular God [§39].

> Rava, and some say Rav Hisda, said; if a person is visited by suffering, that person had better search out his or her conduct (Brakhot, 5a).

If a person is miserable, it must be that person's fault. Rava puts the blame on us citing a verse from the Book of Lamentations (3:40); one must examine one's ways, and return unto YHWH. Does that mean that all the happy people are righteous, by definition, and the unhappy people are not? Is that Rava's lesson?

But Rava is not so narrow-minded. We met Rava earlier as a man who intervened in God's natural order and dared to create a *golem* [§27]. Rava was also acquainted with suffering; he was the Rabbi who unintentionally wounded himself in his study of the Torah. Rava assigned Israel's confirmation of the Covenant to the tense days of Ahasuerus in Israel's unhappy Exile [§53]. Rava does not equate happiness with morality. Rava continues his analysis of suffering:

> But, what if the suffering person has made a careful search of his or her deeds and has come up with nothing?

What does Rava mean that the soul-search "has come up with nothing"? Is there a person in this world free of sin? Rashi, commenting on Rava's statement, explains the unproductive search quite simply:

> The search failed to uncover a sin of a magnitude worthy of that degree of suffering.

The punishment must fit the crime.

(I am reminded here of a sermon on hell told by James Joyce in his autobiographical book, *A Portrait of the Artist as a Young Man*. A Jesuit preacher tells young Stephan Dedalus and his classmates that even "a single venial sin, a lie, an angry look" is so "hideous" in the eyes of God that "the misery in the world" is just punishment for human weakness. James Joyce, who left the Church, is not a reliable spokesman for the Jesuit view of the world; I cite Joyce only as a counterpoint to Rava. Even God, says Rava, in the words of Rashi, must be reasonable. Joyce's Jesuit preacher would say that God's management of creation is reasonable to God, although not always to us.)

Rava cites verses from Scripture, which we shall skip, to support two alternative explanations for unreasonable suffering. Explanation number one: If the person can find no serious sin of commission, then perhaps the person has sinned by omission. That person may not have exercised his or her unique human mind.

> The suffering person should blame the suffering on his or her failure to study Torah.

The second explanation for misery is even more intriguing:

> And if failure to study is not a convincing reason for the suffering, then the person should know that the suffering comes of love. As it is written; "For

YHWH corrects the one He loves, just as a caring father corrects his son. (Proverbs 3:12).

Here then is Rava's approach to the "misery of the world": Study the Torah, and through study you will understand the meaning of your suffering (or perhaps the suffering will lose its sting). If study does not alleviate your pain, then know that your pain is a specific sign of God's love for you His son.

> Rava said in the name of Rav Sehorah who said in the name of Rav Huna; The Holy One Blessed is He afflicts with suffering the individual He most esteems. As it is written (Isaiah 53:10); It pleases YHWH to afflict him.

But, continues Rava, unjustified suffering signifies God's love only when the sufferer loves God in return, despite the suffering. The objective test of God's love is the person's love of God. Indeed, love of God despite one's misery is consistent with Rava's reasoning about the "days of Ahasuerus" [§53]. The Covenant between Israel and God is confirmed only when Israel adheres to God's Covenant in times of trial; the love of God is expressed best when God is loved specifically at moments of pain. Love of God in health and well-being is no test of love.)

Rava goes on to cite a verse from Isaiah (Isaiah 53:10) that likens a person's pain to a sacrifice. And Rava continues,

> Just as a sacrifice brought to the Temple requires the full intent of the person who brings the sacrifice, so must a person's undeserved suffering be willfully accepted by the person.

Undeserved suffering is a sign of God's love only if the person has received the pain lovingly, as if the pain were a personal sacrifice brought by that person to God. Undeserved suffering is a sacrament. Undeserved suffering can be a signifier of

God's special attention, only when the person persists in loving God.

Note the cycle of mutually dependent signifiers: a person's love of God, despite that person's misery in this world, is a sign that God loves that person and has bestowed the misery as a signifier of His love.

The Bible explores undeserved suffering in the Book of Job. The Talmud's analysis of undeserved suffering, however, is more enigmatic than is the prototypic suffering of Biblical Job. Job's misfortunes are a test of fortitude instigated by Satan (Job 1:1-12). Job's pain is not a sign of God's love of Job, but of Job's strength of character.

The Talmud, faithful to its case-history method of analysis, proceeds to probe the sacrifice metaphor (Brakhot, 5a). How can an objective observer judge whether the suffering person accepts his or her pain with love for God? The Talmud likes to see empirical evidence, not mere claims. How can we be sure of the sufferer's love for God despite misery? Here is the evidence: If despite the pain, the person continues to study Torah or to pray to God, then we can be sure that the pain is a pact of love (Brakhot, 5b).

This assertion is circular. We can infer something about God's intent by observing the individual's response to undeserved physical or emotional suffering. God would not rob a beloved person of the pleasure of the intellect or of the comfort of prayer. This circular argument becomes a vortex: A person who gives up study and prayer is one who does not love God. The circular relationship, in fact, grants the individual cosmic power; the person who does love God in misery (the person persists in study and prayer) proves that the misery is a gift of God; the sufferer thus compels the love of God in return for the person's love of God.

Connecting undeserved pain with sacrifice, atonement, and a bond of love between a person and God might seem surprisingly close to a Christian point of view. The surprise lessens when we recall that Christianity grew out of certain trends of Jewish thought; this treatment of suffering in Brakhot reflects the common root of Christianity and Rabbinic Judaism. But, the Talmud prefers practice to theory, and leads us back to this world. Brakhot, 5b, faithful to the empirical method of the Talmud, continues with case histories:

> Rabbi Hiya son of Abba fell ill. Rabbi Yohanan came to comfort him. Yohanan said to Hiya; Is your suffering dear to you?
> Said Hiya; Neither the suffering nor its reward.
> Said Yohanan; Give me your hand.
> Hiya gave him his hand, and Yohanan made him rise healthy.

> Rabbi Yohanan in his turn fell ill. Rabbi Hanina came to comfort him.
> Hanina said to Yohanan; Is your suffering dear to you?
> Said Yohanan; Neither the suffering nor its reward.
> Said Hanina; Give me your hand.
> Yohanan gave him his hand, and Hanina made him rise healthy.

> Why did not Yohanan cure himself?

We just saw Yohanan cure Hiya.

> The Rabbis said; The lone prisoner cannot free himself from the prison cell.

No man can free himself because help must be mutual; help comes through interaction with others.

> Rabbi Elazar fell ill. Rabbi Yohanan came to comfort him, and saw the sick man laying in darkness.
> Yohanan bared his arm, and the room was filled with light.

Yohanan was a man of such lustrous beauty that his very flesh glowed in supernatural light. The Talmud (Baba Metzia, 84a) likens the glow of Yohanan's beauty to that of a silver cup filled with red pomegranate seeds in a bed of rose petals placed in the interface between sunlight and shadow—a metaphor of life science [the rose], material science [the silver] and physics [the light]. Yohanan's beauty turned shadow into light.

> Yohanan then saw that Elazar was weeping.
> Said Yohanan; Why do you weep? If you weep because you did not study sufficient Torah, know that the measure of study is one's intention to study; if you weep because of poverty, know that not all are destined to be satisfied; if you weep because of your children who died before you; see this bone of my tenth son.

Yohanan himself suffered the death of ten children, and carried a bone of the youngest as a sign of his commiseration with others in grief.

> Said Elazar to Yohanan; For none of this do I weep. I weep for your beauty which is destined to become dust.

Yohanan has come to commiserate with Elazar, who is his student, and Elazar is he who commiserates with Yohanan, his teacher, and with the beauty of the individual world that is condemned to death.

> Said Yohanan to Elazar: For that you are surely right to weep.
> And both wept together.

The comforter and the comforted, the student and the teacher, weep together for the tragedy of death that befalls both the righteous and the wicked. We have come full circle to Rabbi Abbahu who valued the rain above the resurrection [§13].

By and by, said Yohanan to Elazar; Is your suffering dear to you?
Said Elazar; Neither the suffering nor its reward.
Said Yohanan; Give me your hand.
Elazar gave him his hand, and Yohanan made him rise healthy.

The Talmud faces the tragedy of life and death head on. No inherited "Original Sin" makes us born guilty and no other-worldly "Paradise" frees God of responsibility for death. We are back to Rosh Hashanah and its covenant of acceptance [§60]. The Talmud's answer is to give a hand and take a hand and help one another rise out of misery to go on with life in this world. Baba Metzia, 84a, tells us how a tragic death came to the once beautiful Yohanan to relieve him of his final pain and madness. But Yohanan of the Talmud still lives and acts in the text.

Brakhot, 5b, closes its journey into suffering with a flash of ironic Jewish humor—a Talmudic forerunner of the Jewish stand-up. The vignette is a parody of the suffering of Job. Job, in the Biblical story, is tested in the loss of his property, his servants and his children in a series of disasters (Job 1:13-20), and Job is afflicted physically (Job 2:7). Job's friends come to comfort him (Job 2:11-13). Rav Huna suffers disaster on a different scale, but his friends too come to comfort him:

> Rav Huna was afflicted by the loss of 400 casks of wine that spoiled by fermenting into vinegar. His friends came to comfort him: Rav Yudah the brother of Rav Sala the Pious and other Rabbis, and some say the comforters were Rav Ada son of Ahava and other Rabbis.
> And they said to Huna (in simple Aramaic, not in the elevated Hebrew of Job); Let the master search out his deeds.

There must be a reason for such a serious loss of wine.

Said Huna to the comforters; Do you suspect me (of sin)?
Said the comforters; Are we to suspect The Holy One Blessed is He of injustice?
Said Huna to the comforters; OK, if anyone here has something to say about me, let him stand up and say it openly!
Said the comforters; We have heard rumors that you do not let your tenant farmer take the pruned cuttings of the vines (which the Halakhah has ruled belong to the tenant farmer).
Said Huna; What are you talking about giving him the cuttings? This fellow has been robbing me of my grapes!
Said the comforters; You know the saying—one who acts like a thief is a thief.
Said Huna; OK, OK, I'll give him the cuttings.

Huna admits that he is in the wrong with his tenant farmer and that the punishment is deserved and fitting. Huna is quoted by Rava above as one who taught that God "afflicts with suffering the individual He most esteems". Yet, Huna accepts his suffering as a sign of fault, not of esteem.

Steadfast Job is rewarded at the end of the Bible story by even more property and by the birth of new children (Job 42:10-17). Huna too enjoys restoration.

Some say that the vinegar was transformed back into wine. And some say that the market price of vinegar rose to equal that of wine.

§67 Talmudic hermeneutics

Our meander through Talmudic texts, which is now ending, is an exercise in hermeneutics. Hermeneutics is the term that refers to the art of interpretation. The word hermeneutics originates from the Greek, and would seem to be derived from the name of the god Hermes; Hermes was (or is) the messenger of the gods and himself the god of science, commerce and language (Oxford English Dictionary, Second

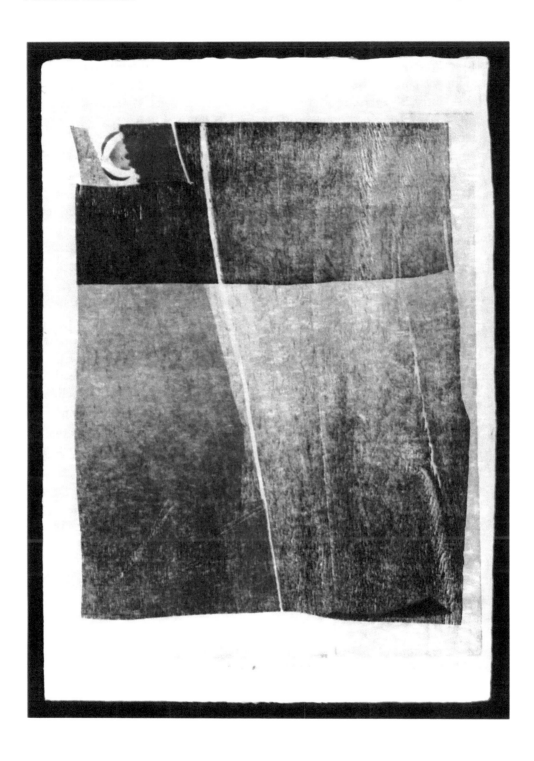

Edition, 1989). Thus the connection between science and interpretation is classically antique and encompasses language and the privileged communication of hidden information by adepts such as Hermes. A scientist too may be perceived as adept with hidden information. Like Hermes, a scientist is a messenger who discovers and reveals information about nature. But before we discuss the hermeneutics of science, let us characterize the hermeneutics of the Talmud.

A defining illustration of Talmudic interpretation can be found in Eruvin, 13b.

> Rabbi Abba said in the name of Shmuel; For three years the House of Shamai and the House of Hillel were at odds. This side said, the Halakhah is as we interpret it, and this side said, the Halakhah is as we interpret it.

Shamai and Hillel were two sages who established fiercely opposing (Houses) schools of thought regarding almost every aspect of the *Halakhah*. The controversy took place at the end of the first Century BCE, a formative moment in the development of Talmudic Judaism. The text in Eruvin, 13b, tells us how the conflict finally was resolved:

> A Voice from Heaven (bat kol [§21]) was heard to say; Both your opposing sets of interpretation are the living words of God, and the Halakhah is according to the House of Hillel.

Unlike the episode of Akhnai's oven [§21], on this occasion interference from Heaven is not rejected by the Rabbis; Heaven, in this instance, apparently expressed the consensus. But ought not we object to the Voice? How can two mutually exclusive interpretations of the Torah both be the *living words of God*? Reason would argue that if the House of Hillel asserts the truth, then the opposing interpretations of the House of

Shamai must be false. But the Voice from Heaven is careful to make a distinction between interpretation and *Halakhah*; it's only the *Halakhah*, the code of behavior, that must be resolved unequivocally. A society cannot sanction contradictory behavior; one law must hold for all. The law, announces the Voice, is to be carried out according to the House of Hillel.

Differing ideas and opinions, however, can be tolerated. The Heavenly Voice tells us that mutually opposing interpretations do not necessarily cancel each other's value. The episode described in Eruvin, 13b, in fact, tells us that the Talmud not only tolerates multiple interpretations, it values and cultivates them.

Note that the Heavenly Voice described both interpretations as the *living* words of God. In what sense does the controversy between Hillel and Shamai make the words of God *live;* are not the words of God sufficiently vital without human discussion? One might answer that the words of God gain an added measure of life because of the passion with which Hillel and Shamai continue to discuss them. The words of God, in fact, materialize only in their interpretations [§54]. The Talmud in Eruvin, 13b, interprets Heaven's choice:

> If the words of both Houses are truly the living words of God, why was the House of Hillel chosen to formulate the Halakhah?

The Talmud gives three reasons:

> Because the sages of the House of Hillel were patient and meek,
> Because they studied the ideas of the House of Shamai and not only their own ideas,
> And above all, because they mentioned the opinions of Shamai before their own.
> All of which comes to teach us that whoever lowers himself is raised by God, and whoever elevates himself is put down by God.

The preference for Hillel might have been established on theological grounds ("the opinions of Hillel are closer to what God had in mind"); or on political grounds ("the people support Hillel"); or on rational grounds ("Hillel is more reasonable"); or for social reasons ("Hillel is with the downtrodden and Shamai is elitist"). But the Talmud subordinates the theological, the political, and the social to the ethical; remember how Hillel summarized the Torah: *That which is hateful when done to you, do not do to your fellow* [§42]. Hillel won acclaim because he and his followers were meek, retiring and modest. Be a better person, and reality will respond [§66]. The incident described in Eruvin, 13b, certainly invites interpretation; the point here is that the Talmud (and Heaven) endorse differing interpretations. Interpretation is not either true or false. A continuing process of interpretation, even conflicting interpretation enlivens the word of God.

Obviously, however, there must be limits to Talmudic interpretation. Not every idea that comes to mind is acceptable. What constraints keep interpretations from getting out of hand? Quite simply, the process of interpretation is constrained by the rules of interpretation. The rules of Talmudic interpretation are very complex and we cannot discuss them here. Suffice it to say that the Rabbis underwent rigorous training to learn the tradition and the procedures of interpretation. Interpretation, nevertheless, is not determined by authority. A knowledgeable and skillful interpreter is free to interpret. He or she need only convince the peer group [§21, §28].

Actually, to do proper science, a scientist too has to acquire basic information and become adept at scientific experimentation and methods. The rules of science mould scientific creativity, just as the rules of Talmudic interpretation mould rabbinic creativity. A scientist becomes a scientist by learning the rules of the scientific enterprise. The scientist too is licensed to make new interpretations; as we shall see, he or she need only convince the peer group to validate them [§78].

To approach the rules of Talmudic interpretation, you can study in a Yeshiva [§11] or read various books. *The Essential Talmud*, by Adin Steinsaltz, cited above [§6] is a good place to start. For those interested in literary criticism, insight into Talmudic interpretation can be found in *The Slayers of Moses. The Emergence of Rabbinic Interpretation in Modern Literary Theory*, by Susan A. Handelman, State University of New York Press, Albany, 1982.

PART III

Texts of Science

This section presents the idea that science, like the Talmud, depends on skillful interpretation of a received text. The text of science is nature, and science reads and writes this text according to paradigmatic rules. We discuss interpretation, understanding, signification, meaning, communication and data. We explore the idea of free will and compare the interpretation of meaning done by systems of cells with that done by collectives of people.

§68 Interpretation

Scientists, like Talmudists, interpret; the difference between them lies in the subject of interpretation. Scientists interpret the results of experiments. Now, it is generally thought that experimental results need far less interpretation than do Talmudic texts. The results of experiments are believed to speak for themselves; any informed scientist who looks at the data is expected to arrive at the same conclusions. Scientific interpretation is supposed to be transparent. Or so most scientists (and non-scientists) would like to believe. The point here will be that interpretation is central to science too. The bare face of nature, as we shall see, requires cosmetic interpretation before it can be noticed. A scientist has to interpret nature in order to note her essential beauty [§85]. But let us not jump ahead of our story.

What is interpretation? To interpret is to decipher and explain a hidden meaning. The manifest subject of the interpretation—be it an object, a story, a process, or a text—is seen to encode some hidden entity—a different object, story, process, or text. To interpret the manifest subject is to break the code and reveal the covert entity. The covert entity is thus signified by the manifest entity; to decipher the hidden meaning of a text is to explain its signification. The Talmud, as we saw above [§67], can accept the validity of opposing interpretations. What about science; is there only one true interpretation nature? Does science achieve an exclusive truth? Is scientific interpretation only simple decoding? Keep that question in mind as we proceed.

Our keywords for analyzing interpretation will be: *understanding, signification, meaning,* and *communication.*

Understanding

Understanding involves metaphor, prediction and know-how. To understand a thing is to be proficient in interacting with it. Understanding causality, complexity and free will challenges both science and the Talmud.

§69 Understanding by metaphor

What does it mean to say that you or I understand something? What is the essence of human understanding? We demonstrate understanding, I suggest, in three ways: we visualize *metaphors*; we make successful *predictions* and we exercise know-how—*utility*. In other words, we show that we understand by grasping likenesses, by foreseeing consequences, and by knowing use. Understanding, from this point of view, is not a state of mind; understanding is a type of performance; understanding is interacting with proficiency. Let us start with metaphor.

A metaphor is a name or description that is transferred to some object different from, but analogous to, the object to which it is usually applied. Seeing an analogy between seemingly unrelated ideas is the key to making a metaphor. You construct metaphors by interacting productively with ideas. The human mind understands through metaphors; the mind understands when it sees that one thing is really like another. We understand the *strange* and the *puzzling*, for example, by seeing their analogy to the *familiar* and the *known*:

> Men believe they understand women when they imagine that women really think like men think, with some differences.
> Women think they understand men when they perceive that men are just like children, only bigger and stronger.

We understand the *complex* by seeing that it is like the *simple*:

> The human body is a machine; the eye is a camera; an airplane is a metal bird; the brain is a computer.

Understanding amounts to translating one thing (the unknown) into the terms of another (a known). The unknown is *mapped* onto the known, and so becomes known. Metaphorical mapping comes about by fitting patterns. A road map, for example, helps you find your way because the patterns of the lines drawn on the paper fit the patterns of the actual roads crisscrossing the segment of the earth that is represented by the map; the road map is a functional metaphor for one aspect (the roads) of a part of the world.

How is it that your brain can see that a map held in your hand is true to the patterns of the actual roads? You see the roads on the map all at once, but the intersections you drive by in the real world are separated by intervals of time. The real roads are seen on a vastly larger scale than the mapped roads, which are only ink on paper. How do you manage to associate the map with the reality of the reads? How can you tell that the map is a map at all? How, in fact, do you manage to find your car in an unfamiliar parking lot? We don't know how the cells in our brains create and consult such maps, but humans are masters in the art of representation. Representation is our specialty. Maps, pictures, stories, songs, and words are living examples of our aptitude for encoding reality in our heads. A metaphor can be viewed as a type of association.

All metaphors are abstractions. It is true that a road map is no less concrete than are the actual roads the map depicts; both materially exist. Nevertheless, the analogy you see between the road map and the roads is an abstraction.

UNDERSTANDING

The concept of analogy, like any concept, is abstract. Thus, all analogies and all metaphors, even those between concrete objects, are abstractions made by human minds.

Metaphors, we may say, are ways of encoding one abstraction in the terms of a different but analogous abstraction. We understand the meaning of Orpah's name, for example, when we see that the meaning of the word *orpah* encodes the girl's behavior when she turned her *back* on her mother-in-law Naomi [§52]. A metaphorical analogy may encode an association of occurrences. Rabbi Eliezer, for example, associates the reciting of the *Shma* with the Sabbath because both ceremonies occur at nightfall [§41]. Metaphors arise as abstract associations between the overt and the covert. The more fitting the association, the more convincing the metaphor, and the greater the understanding. Metaphorical associations satisfy the human mind. Once we find a convincing metaphor, we find pleasure. Understanding—a good metaphor—comforts.

Why do we like to make associations and construct metaphors? Consider this: A creature can be said to be adapted to its environment when its body structure and its patterns of behavior fit the character of its niche in the world. Fur, claws, hooves, wings, shells, scales, gills, fins, horns, color of skin or feathers and other body structures are adapted to particular life styles. Particular programs (instincts) for display, killing, fleeing, eating, sleeping, mating, nesting, nursing and other activities are behaviors adapted to particular life styles. The adaptations of a species to its world can be seen as a type of representation by the creature of the environment within which it lives: wings represent the properties of the air used for flight; lungs represent the properties of the air used for breathing; gills map the properties of water that facilitate the extraction of oxygen; the claws and hunting patterns of lions represent the structure and behavior of zebras; the present characteristics of wheat, maize and rice encode the preferences of humans, and so forth. The successful creature encodes in its structure and behavior the piece of reality that houses it.

Most species are born well adapted by evolution to their worlds. Goldfish, trees, houseflies, worms, bacteria and other creatures are born hard-wired to their environments; they need not bother much with learning; they know what to do automatically. More complex creatures must invest more time learning. Dogs, cats, lions, horses, goats and humans among others have to practice mapping their environments and improving their behaviors. Puppies, kittens, cubs, foals, and kids learn what they need to learn by exploring and acting. The exploratory activities of immature organisms we call *play*. Animals that must learn a way to survive in their world do so by playing: the developing puppy, kit and cub play at stalking; the foal and the kid play at fleeing. Social species play at communication, cooperation and dominance. Play allows a creature to internalize its environment in preparation for life in a particular niche. Developmental play-time is adjusted to the amount of learning the creature needs. Fish, trees, flies, worms and bacteria play not at all. Puppies, kits and cubs go through a stage of play as they grow and mature. Humans, who have the most to learn, require the most play.

Humans are born physically ill-equipped. No wings, no claws, no speed, no fur, no shell for survival; the mind and the hand are all we have. To survive, we each must spend a lifetime adapting ourselves to a changing world. Each has to adapt to a unique world because each creates a unique world. Individuality is built into the human species [§1]; see my book, *Tending Adam's Garden:*

Evolving the Cognitive Immune Self, Academic Press, San Diego, 2000. We have the most to learn, so we humans have evolved to embody perpetual curiosity. We play with everything—ideas, objects, people—until we learn to understand them. When we are too tired, too inept, or too timid to play ourselves, we watch others play on the field, the ring, the stage, or the screen. And we learn to understand ideas, objects or people by discovering (or inventing) the likeness of one to another. Much of learning is constructing associations. We transform the strange and threatening into the familiar and useful by discovering their likeness; we associate one with the other. So a good metaphor comforts. The opposite is also true; an unexpected metaphor can transform the tritely familiar into a shocking delight; a discomforting metaphor can also enlarge understanding. Our evolutionary fitness to our niche in the world comes about by having an open mind and a lot of curiosity. Bees are built to hunt flowers, lions to hunt zebras, and humans to hunt metaphors. We survive by metaphor making.

But, you may argue, a metaphorical definition of understanding fits poetry, not science. Metaphors are literary devices, while science deals with the reality behind the appearances of nature. Science seeks quantitative understanding. Science rests on mathematical precision. So how can scientific understanding be merely metaphorical?

The answer is that mathematical descriptions of nature are no more than metaphors. Mathematics, however precise, is only a language. The equation describing the law of gravity is not gravitational attraction itself; the equation describing the law of gravity is a metaphor for a *regularity* in the behavior of mass; two masses will interact in a way that fits the equation—the mathematical metaphor. Science aims for the fitting equation. The equation is a translation of a regularity of nature into a language familiar to (some) human minds. No less and no more. Indeed, "pure" mathematics, like any other language, can provide pleasure to human minds without any noticeable connection to nature. Nature (as she exists outside of the human mind) has no knowledge of mathematics, whether formulated by Einstein or by accident; she just does her thing. The human mind recognizes the regularities of nature and supplies precise mathematical metaphors.

Not all human minds can supply the math, but all can supply the words. Words too are metaphors. We have words that we associate with objects, with actions and with interactions. Language, a string of words, is our ultimate metaphor for reality. Recall that the Hebrew word *davar* means both a *word* and a *thing* [§9]. Indeed, the teacher tests the student's understanding by asking him or her to answer a question—to provide a description (a translation) of the test object "in their own words". Understanding involves a proficient use of language, math or other metaphor. Competence in language is akin to competence in metaphors, which is akin to competence in mapping. Language is the uniquely human map.

We are comfortable with words. We feel secure when we command a language that suits the situation. The naming alone of an object (person, concept, molecule, process) can convince us that we understand sufficiently the object (person, concept, molecule, process). Verbal aptitude might be a reason why scientific experience can be the enemy of scientific creativity; the veteran scientist may come to believe that he or she has sufficient understanding when he or she merely has stereotyped words and worn metaphors. Creativity is not driven by understanding. Creativity is driven by curiosity—the continuing search for new metaphors.

Creativity is a unique characteristic of the human brain. Computers are far superior to us in memory and can perform complex computations with relative ease; we, in contrast, forget names and menace the roadway when our attention is diverted to a mobile telephone. But no computer can match the human brain in making productive associations. Thoughts trigger thoughts as we brainstorm our way through life; smells trigger visions; encounters trigger dreams; sounds call forth concepts. Metaphors are associations, and so metaphors are not merely figures of speech or ersatz understanding. Metaphors are true understanding—competence in metaphor making is the very stuff of human intelligence.

(I have used the term *metaphor* here in a loose way; to see how the term is used by professionals, you can read *Metaphor and Thought, Second Edition*, edited by Andrew Ortony, Cambridge University Press, Cambridge, UK, 1993.)

§70 Prediction and understanding

The second test of adequate understanding is successful prediction. We demonstrate that we understand an object or process when we are proficient in forecasting its future behavior. Detecting a regularity of nature with scientific precision empowers us to make scientific predictions. The regularities we discovered about gravity and motion allowed us to send a man to the moon. The regularities we discovered about material interactions allowed us to formulate the periodic table of elements and to do chemistry. A scientist's ability to make a scientific prediction serves as a test of adequate understanding. The regularities we have discovered about the behaviors of spouses, children, and parents give us the feeling that we understand them. (Until we discover otherwise.)

§71 Utility and understanding

We also understand a thing when we know how to use it. Proficiency in use is the third test of understanding. Usefulness is related to prediction. We understand an illness sufficiently when show we know how to diagnose it and cure it. We demonstrate that we understand particular laws of physics when we make an atomic bomb or send a man to the moon. Technology is the utility of scientific understanding.

§72 Understanding causes

Science strives to understand causality: that which makes nature act the way she acts. One can divide questions of causality into two types: we can wonder about the causes of this or that particular event (Why is it raining today?), or we can ask about the causes of a general class of events (Why does it rain; what are the rules governing rainfall?). The causes of particular events occupy courts of law (Who done it?), engineers (Will this building withstand the next earthquake?), artists (A Portrait of the Artist as a Young Man), or parents (Why don't you learn to behave?). A social worker may need to understand why a particular baby was battered by his parents, and the baby's doctor might want to know what medication to prescribe for that child's attention deficit. Particular causes can usually be assigned to particular causal agents—the powers or persons that can be praised or blamed for the event. The Talmud devotes much discussion to causal agents, and not only to general principles. Judicial systems in particular function to determine causal responsibility and blame. Recall how the Talmudic court cautions the witnesses to a murder; direct observation takes precedence over circumstantial interpretation [§14].

But the biologist wants to understand the nature of reality in general; not what happened to this or that baby, but, for example, the rules and processes by which the genetic program makes fertilized eggs develop into babies. Scientists, as professionals, are interested in general causality. Science seeks the relevant laws of nature, not the fortune or misfortune of the individual exemplar. Obviously, the more we understand the general properties of living organisms, the more help we can extend to this or that individual or class of individuals. The more we learn about solid-state physics, the better our electronic devices will perform. Particular applications emerge from understanding general causes. So what does it mean to understand a general cause?

Understanding general causality is akin to discovering how an overt piece of reality is encoded in a covert piece of reality. For example, the movement of the earth and the other planets around the sun (the overt reality) is encoded in the laws of gravity and motion (the covert reality). Or human speech (the overt reality) is encoded in the way the brain works (the covert reality). Usually, the overt portion of reality is *macroscopic*, in the sense that we can see it or feel it. The covert portion of reality responsible for macroscopic reality is often at a smaller scale; the covert cause is *microscopic*. We need special equipment to see it; we need a microscope to see a cell; we need imagination to see an atom. Scientific understanding, for the most part, amounts to uncovering microscopic entities (cells, molecules, atoms, energy, processes and information) that encode (underlie) the macroscopic world of our experience. Understanding causality emerges from discovering the regularities at the microscopic level of nature that match the macroscopic regularities of the world we experience.

Nevertheless, not all macroscopic regularities are caused by covert regularities at a smaller scale: for example, sunrise and sunset, clearly macroscopic realities, are caused by the organization of the solar system, a covert *megascopic* reality. Indeed, the solar system is influenced by the galaxy, an even larger covert reality. However, the organization of the galaxy and the solar system is caused by the laws of gravity and motion, and these are microscopic. So the covert realities that cause overt realities may be found at both smaller and larger scales.

Note an important distinction between causality and our understanding of causality. Ideally, we have been taught to seek a single causal agent for every effect. We use the metaphor of a *causal chain*; causes, like beads on string, are imagined to be discretely connected one to another in sequence. The *domino effect* is another metaphor for causality. The image here is of row of dominos standing on end. Pushing over the first domino leads to a chain reaction in which they all fall in sequence, each pushing its next neighbor as it falls. Sequential causal chains are simple to visualize, and this simplicity leads us to believe that causal relationships are objective; the identity of a causal agent in a chain reaction should be independent of one's point of view.

The problem is that causality in the arena of nature is more a complex web of circumstances than a discrete chain of events. Consider a car accident: One might say the accident was caused by the fact that both vehicles entered the intersection at the same time. However, what about the fact that one of the drivers had imbibed a beer? What about the fact that one of the cars had faulty breaks? What about the poor visibility? One's understanding of the accident would seem to depend on one's particular interests and point of view. We understand causality by making metaphorical representations ("maps") of events. Individuals make different maps, and so individuals vary in their ideas of what constitutes a cause. There is room to interpret causes because causality, as we have said, is not a fixed chain of single reactions. Natural causality is a web of mutual interactions between the regularities of nature at many scales. The discovery of causes is the discovery of patterns of interaction that are associated with other patterns. The causal agent, in this context, is not a discrete force, but the mutual influences of patterns of interactions.

Consider the following: Patterns of forces and elementary particles interacting in exploding stars generate light, heat and atoms; patterns of atoms interact to form patterns of molecules; patterns of molecules interact to form planets and the earth; and patterns of molecules on earth interact with the help of the sun's energy to create life and the biosphere. The pattern of evolution of life on earth generates the development of species that interact with each other in patterned ways: lions interact with zebras and zebras interact with grass and humans and bacteria interact with all the world. Inside the cell, the patterns of genes interact with the patterns of proteins and vice versa. Patterns of interactions between nerve cells create the human brain and its ideas; patterns of ideas generate history and metaphors and interpretations; and patterns of interpretation produce the Talmud and science and the rest of human culture. The discovery of how *patterns* of being mutually *interact* with certain *regularity* is the discovery of the fabric of causality. And, as we discussed in the beginning [§5], a meaningful fabric is a text. So we can justly claim that science reads and interprets the texts of nature.

(For a discussion of causality in biology, see Irun R. Cohen and Henri Atlan, Limits to Genetic Explanations Impose Limits on the Human Genome Project, in *Encyclopedia of the Human Genome*, Nature Publishing Group, Macmillan Publishers Ltd, 2002.)

The Talmud too has its own causal point of view. The text of the Talmud, in contrast to the texts of science, is not concerned with the exchanges of matter, energy and information that mould nature. The Talmud relates patterns of causality in the world to mutual interactions between humans, nature and God [§66]. The Talmud attributes the state of the world to causal webs of moral values and obligations, individual and collective, that mould history. The Talmud, as do most other religions and social systems, holds people responsible for their behavior; the Talmud endorses a concept of free will.

§73 Emergence of free will

People often think before acting. This connection between our thoughts and our deeds gives us the impression that our actions can result from our deliberate thoughts. The process of linking action to thought we call *choice*, or *free will*. Although we exercise our will freely every day, we are not quite sure how we do it. The idea of free will would seem to be at variance with the concept of causality developed by science. What could be the meaning of free will in a universe where every event has a cause? How might a person ever willfully make a decision that is not imposed by outside forces or internal compulsions?

The idea of free will would also seem to be at variance with the concept of the all-knowing God taught by monotheism. What could be the meaning of free will in a universe where God knows everything that was, is and will be? How might a person ever make a decision that is not already foreseen? Wherein lies our freedom? Thus the concept of free will is problematic both for science and for the Talmud.

Let us begin with the scientific problem. The scientific challenge to the idea of free will arises from our view of causality, particularly the causality seen in the classical mechanics described by Newton's laws. To move a resting mass or change the movement of a moving mass, for example, requires a force. To open a door requires pushing or pulling, as the case may be (some doors slide and are pushed by motors). A force transmitted by the cue pushes the billiard ball into the side pocket. Classical mechanics describes the domino effect, how a falling domino produces a chain reaction [§72]. These mundane events are standard metaphors for discrete causality. Now, if we view a person's choice of action in the same way we view a falling domino , a rolling billiard ball, or an opening door, then we can logically conclude that the person's action too must be caused by a discrete force, or a chain of discrete forces. If we can reduce the course of the billiard ball or the fall of the domino to the imposition of causal forces, then we should be able to reduce human behavior to the imposition of causal forces. True, your thoughts move your body; but your thoughts too must be motivated by chains of causal agents. Are we only puppets activated by strings of external causes? According to classical Newtonian mechanics, the answer would be yes.

The causal descriptions of classical Newtonian mechanics, however, do not adequately describe all aspects of the physical world. Some scales of material reality are not Newtonian. Phenomena in quantum mechanics, radioactivity, and other areas of physics do not seem to be ruled by discrete causes, but rather appear to be motivated by statistical, or even chance considerations. It is not our business here to enter into the details; for more information on the subject see, for example, *In Search of Schrodinger's Cat: Quantum Physics and Reality*, by John R. Gribbin, Bantam,

Doubleday, Bell, 1985; and *Does God Play Dice? The Mathematics of Chaos*, by Ian Stuart, Penguin Books, 1990. I leave to physicists the questions of quantum causality. I raise the issue here only to illustrate that different views of causality suit different scales of reality. Certainly, the willful vagaries of human behavior are not explained by quantum mechanics. Human behavior and quantum mechanics operate at very different scales of nature. The human will, free or not, is definitely not explained by Newtonian mechanics. As we discussed above [§72], causality is more a web of circumstances than a chain of discrete events. The human will is related, I believe, to a much higher scale of complexity; the will is an *emergent property* of the human brain.

A complex system such as the human brain cannot be understood by cataloguing its component parts. We already know the components of the brain. The brain is made of brain cells and their connections. Yet knowing all there is to know about the physics of nerve conduction, the biology and biochemistry of brain cells, and the anatomy of nervous connections will not explain comprehension, speech, or consciousness. Each cell just produces electrical impulses and chemical substances in response to other electrical or chemical signals. The properties of mind are not electrical or chemical signals; the mind *emerges* from the collective patterns of interaction of an enormous number of brain cells. The brain shows us that patterns of interaction can create a new level of reality—the mind. The mind is not a mysteriously independent entity; the mind is an emergent property generated by the complex interactions of the material brain. The very complexity of the interactions of millions of brain cells generates qualities distinctly beyond those of any of the component cells of the system. The system, in short, cannot be *reduced* to a catalogue of its parts. The historic mind-body problem arises from the misguided attempt to reduce the complexity of the brain to the laws of physics that govern its component parts; it simply cannot be done. The complex system, through its complex interactions, outdoes its parts. Co-respondence in the immune system [§79], economies, political systems, living organisms, the evolution of life, are all examples of the emergence of new aspects of reality arising from complex patterns of interaction [71]. Complexity creates. Life, as we all know, is just a further evolution in the complexity of carbon atoms. How are we to understand the causes of emergence in complex systems?

As we discussed above [§68], understanding is based on representations—words, pictures, mathematical formulations or computer simulations. To understand with scientific precision, we have to formulate a precise mathematical metaphor of the system and the way it behaves. Can we do that for a system as complex as the mind?

A human mind makes willful decisions by consulting with itself: Do I want to study medicine or law? Do I want to marry this person or that person? Do I prefer the red wine or the white wine? These internal deliberations are influenced not only by the profession, the person or the wine facing us, but by our past history, memories, the recent state of our brain, our tastes, genes, education, social identification, and group dynamics. The complexity of patterns of brain behavior is, at least for now, beyond adequate scientific treatment. There is simply no existing computer or computer program that could faithfully model the actual complexity of the decision-making process. Understanding is limited to conjecture. The mind of one human can make reasonable guesses about what a fellow human will do; a computer can make statistical predictions. But

these predictions are not precise. Indeed, the human brain is a system so complex that no other system can adequately model it. The only operative model of a particular human brain is that brain itself. The uniqueness of a brain cannot be simulated; brains are one-of-a-kind [§1]. Although every single reaction and interaction within a brain is causally determined at the chemical level, the *will* that emerges from the system is not determinate. We simply cannot account for all the factors that go into the thinking process. What we cannot model in precisely quantitative terms, we cannot understand in precisely quantitative terms. We have to make do with words. The property we call *will* is an emergent property of matter at the pinnacle of its interactive complexity. We use the term *free* as a metaphor for the indeterminacy of this complexity. Inexplicably, consciousness exists; every man and woman knows it. Free will is an emergent property of consciousness.

Of course, one may claim, rightfully I think, that I am using the word *emergence* without a rigorous definition of its meaning. The emergence of an economy through the activities of autonomous people is not the same as the emergence of consciousness from the activities of neurons in a person's brain. In fact, we understand nothing about the neurological basis of consciousness—how the activities of brain cells generate consciousness. To explain free will as an emergent property of consciousness is not a scientific explanation; *consciousness* and *emergence* are themselves poorly understood words. So what is the point here?

My point is that *consciousness* is an empirical fact of human experience despite the lack of an adequate scientific model to explain it. *Emergence* too is an empirical product of complex systems; despite our lack of a complete model of emergence, complex systems do generate attributes beyond the properties of any of their component parts. Hence, *consciousness* and *emergence* are not empty words; although poorly understood, they still exist as common observations. The reality described by these words does not disappear just because that reality defies modeling. *Free will* defies both logic and modeling. But does that shortcoming invalidate our common experience? The complexity of the human brain and its ability to think for itself generate the decision-making process we call free will. That is the point.

Now what about the Talmud's problem with free will? Rabbi Akiva puts it this way:

> Everything is foreseen, and permission is granted (Pirke Avot, Chapter 3:19).

This statement is terse; the Hebrew consists of only four words. The first two words in the sentence say, according to the traditional translation, that God foresees all; God knows what must happen. The second two words state that, despite God's knowledge, people are still allowed to choose between right and wrong. The traditional translation of Akiva's statement is paradoxical. How free could your choice of action really be if God knows before you choose what your choice will be?

The paradox, however, can be resolved linguistically: The Hebrew word (*tzafui*) used by Akiva need not be translated as *foreseen;* the word could as well be translated as *seen.* So if we adopt the alternative translation *seen,* Akiva could be interpreted as saying that God sees whatever a person chooses to do, and let's the person go ahead and do it without interfering. The message then avoids the paradox of foreknowledge and becomes a simple warning: you will be judged by the rightness and wrongness of your choices in life.

But this variant translation of Akiva does not answer the *free will* question; does God know

before you choose what your choice will be? If He really knows, then what are His grounds for judging you as if you were free to choose? I see only two reasonable alternatives: 1) God knows our future actions (we are machines), and so He plays a malicious game of judging us as if we were free to choose our way. 2) God does not know what we will choose to do, and so His omniscience is incomplete.

The Talmud (Avodah Zara, 3a) states that

> The Holy One Blessed is He does not play malicious games with His creatures.

Alternative 2) is more likely if we agree with the above statement from Avodah Zara: He is not malicious and His judgment is no game. God, therefore, must have granted us freedom to choose. He, too, does not know how we will choose to act until we have acted. He too wonders what will be. Don't worry about God's incomplete knowledge; He planned it that way. If God had wanted robots, he could have created angels instead of people. People, who choose and interpret, are far more interesting than angels. Indeed, as we have discussed above, existential time itself materializes only through our choices [§38]. A philosopher may claim that existential time is only our illusion; God's time is eternal, so God knows all outside of time. But we and the world do exist in time, and we do willfully choose. If God deals with people, then God enters existential time.

Is such an idea unbecoming? Consider this: Every atom in your body is the debris of exploding stars; does it diminish God's repute if we believe that He has fashioned (evolved) out of carbonaceous molecules and water a creature endowed with conscious choice? The reader is free to make his or her own judgment of the matter.

SIGNIFICATION

Signification

Science uses the controlled experiment, an artificial slice of nature, to study nature as she really is. This strategy of signification is not unlike the art of the cinema.

§74 Signification in science

Science differs from other forms of human endeavor in its quest to attain understanding at all scales of reality. (For a glimpse at the scales of nature, see Philip Morrison, *Powers of Ten*, Scientific American Books, 1982.) Science is not satisfied unless understanding at the observed (macroscopic) scale is anchored in understanding at the underlying (microscopic) and the overlying (megascopic) levels too. It is not sufficient to understand how to make a decision or read a text, we want to understand how the brain works to produce choosing or reading. Science strives to map the overt (choosing, reading) onto the covert (brain action). What is it that happens inside the brain (and outside the brain too) that gives rise to a choice or a text?

We can consider the matching of the scales of the overt to the scales of the covert in terms of signification. Signification we defined as the connection between a signifier and the signified; a sign is formed by the material signifier together with the underlying concept that is signified [§5]. Nevertheless, we can dissect a sign into its components. We can even identify a signifier without knowing what it signifies [§5]. As we discussed, the matzah is a signifier of the holiday of Passover; but what the matzah signifies needs interpretation [§46]. Similarly, our ability to read signifies some microscopic property of our brain; we just need to figure out what precisely goes on in the brain. Science sees the macroscopic features of nature as signifiers in need of signification, microscopic and megascopic. What is reading in terms of brain action? What does sunrise signify about the organization of the solar system? Science interprets signification.

The experimental methodology of science is itself based on signification. An experiment is designed to serve as a signifier of nature. Nature is too complicated, too uncontrolled and too noisy to study just by looking at her; we have to devise experiments. An experiment is the way a human mind seeks to isolate a particular feature of nature so that it can be probed and tweaked in a particular way. An experiment is a construction, an artifact fashioned by a person or group. The result of the experiment, hopes the scientist, will reveal something fundamental or useful about the regularities of nature that underlie or overlie the measured outcome. Thus, an experiment is a signifier; a piece of nature is the signified.

Galileo, for example, constructed an incline and studied the behavior of rolling balls in his laboratory. He believed that this artificial apparatus would serve as a signifier of a body freely falling in the real world outside his laboratory. He then applied his laboratory results to interpret the "laws" of nature that make real falling bodies fall. (It is said that Galileo may have fabricated some of the data to fit his expectations; but that only supports the idea that the signified precedes the signifier [§5].) Cells growing in tissue culture signify something about cells growing in the body; what we learn about artificial cell cultures we then use to try to interpret the underlying causes responsible for the behavior of cells actually growing in the body. The results of the controlled experiment are analyzed to reveal the natural causes of the real world. Thus, the scientific method deals with observed nature as both a signified and a signifier; she is a signifier of covert causes while she is being signified by the experiment. Science advances along a helix of signification.

It is the practice in research laboratories to discuss the significance of the results of experiments. We sit with our cups of tea or coffee (sometimes wine, if the results deserve it), and the student presents his or her work to the group. Naturally, the discussion centers on whether the experimental results answer the question the experiment was designed to explore; did we get a clear or fuzzy answer to our question (or no answer at all). At the end, I often challenge the students (and myself) to consider whether the experimental result in hand might actually provide an answer to a different question, a question other than the one we had in mind when we originally designed the experiment. Is there, we ask, a question that we did not ask at the time that is more interesting or more important then the question we originally asked, to which the result might also apply. Can we now ask a new question that we might be able to answer using the result in hand? The original idea we first set out to explore becomes, in turn, a signifier of a new question that now needs to be discovered. The result of the experiment becomes, as it were, a metaphor for a new question [§69]. Not surprisingly, this Talmud-like exercise often opens our minds to new ideas.

§75 Cinematic science

In the Preamble, I suggested that the structural form of the Talmud might be viewed as cinematic; many Talmudic incidents are structured as visual takes—close-ups and long shots—dialogues, flash-backs, and other devices used by the modern cinema [§7]. Now, let me suggest that there is something cinematic about the experimental method. Unlike the Talmud, which is cinematic in *structure*, the scientific experiment is cinematic in *strategy*.

The cinema as an art form, like other forms of art, attempts to discover and communicate truths about humans and their world. The strategy of the cinema is to isolate a slice of life (the script), whether realistic or fantastic, and probe it, expose it, and play it out as a surrogate for life in the world. The cinematic story is an artifact that, in its screen portrayal, signifies the human condition and exposes its mystery. The scientific experiment, too, is an artifact that, in its laboratory portrayal, signifies nature and exposes some of the hidden causes of macroscopic reality. The strategy of the cinema is the isolated slice of life, that of the experiment is the isolated slice of nature.

One might claim, rightfully, that other forms of art (the theatre, the novel, the poem), and not only the cinema, present artificial slices of life. Nevertheless, the cinema reminds me most of the scientific experiment because it attempts to explore reality more through concrete visualization than through abstract verbal exposition.

Obviously, the goals of a scientific experiment differ from those of the cinema. The experiment aims primarily at objective data and precise understanding. In contrast, the screenplay aims at presenting and arousing feelings: laughter, pleasure, vengeance, tears, sorrow, fear, satisfaction and catharsis. Of course, effective cinema, like science, can also provide cognitive growth and understanding. Similarly, the scientific experiment, like art, may also provide the scientist and his or her audience with emotional fulfillment: pleasure, sorrow, hope or catharsis. But human feelings, along with wealth, fame and honor are incidental to the scientific enterprise. Such motives, like the lusts of Amram, Meir, Akiva and Pleimo [§26], are sublimated into hard work and creativity [§25].

Meaning

We define meaning as emerging from a process of interaction. We discuss how science tests, validates and up-dates the meaning of scientific information and scientific paradigms. Interpretation is essential to the evaluation of meaning. Biological systems create meaning through democratic process.

§76 Types of meaning

Meaning, like the concepts of causality and signification, is thought to refer to a reality hidden behind appearances. Meaning is akin to a metaphor that exposes a hidden content. Meaning, from this point of view, refers to the validity, the accuracy, the authenticity, the verity of the hidden reality. What is the meaning of the patient's fever and rash? Or (to paraphrase the lyrics of a once popular Israeli song, now banned for its blatant chauvinism) what does the lady mean when she says no? We use the word *meaning* every day because we need it every day; meaning is at the root of understanding. Yet, the concept of meaning is not easy to define because the word *meaning* is used loosely and bears various nuances. We refer to the meanings of words and signs, to the meanings intended by people, to the meanings of events. The various nuances of meaning, nevertheless, do express a common property: meaning, in all its forms, emerges through a process of interaction.

The interaction may be between people. One aspect of meaning, for example, is what the lady of the banned song really meant when she said no; another aspect of meaning is the male singer's interpretation of what she meant. The "true" meaning, in this situation, might be obtained by asking the lady for clarification. (We leave aside the Freudian question of whether the lady of the song really knows herself what she meant when she said no.)

But we cannot always ask the lady; she may no longer be available. What, for example, did Ruth mean when she said "thy people shall be my people, and thy God, my God" [§52]? The Talmud understands Ruth's statement as her intention to convert to Judaism. Is that what Ruth really meant? Did Judaism, as developed by the Rabbis a thousand years later, really exist in the days of Ruth? But does that fact make any difference to the meaning of Ruth's statement today? As we discussed above [§28 and §50], do we need to consult Shakespeare to learn the "true" meaning of Hamlet or of Prospero? Of course not; the meaning of Hamlet and of Prospero, like the meaning of Ruth, is what they mean to us now. This then is the other face of meaning: what we understand to be the meaning of what the lady of the song says. Meaning here is not the lady's intention, but our interpretation of her intention. So when a lady (or man) says no, you had better believe it. Meaning is our response to appearances.

According to this line of thought, the meaning of a word is our response to the way we *hear* the word [§9]. Words only bear meaning to those who respond to them. The response to a word, of course, need not be an action, but only a thought. A word's meaning is the interaction between the word and the community of persons who respond to it. A word that is not capable of eliciting a response has no meaning. A word, then, bears no intrinsic meaning. The meaning is supplied by the response to the word. Even the sound of a word can have different meanings to different communities. For example, the sound *he* means *he* to speakers of English, but the sound *he* means *she* to speakers of Hebrew.

Meaning

rain and resurrection

Meaning, in brief, is two sides of an interaction: meaning can be the *intention* of a conscious being who initiates an interaction (Ruth or Shakespeare); meaning of this kind may be hidden or implied. But meaning also is the *response* of a reacting party to an interaction; meaning of this kind may be overt and objective. Meaning of the second kind does not depend on conscious intentions. Meaning of the second kind is the response to a stimulus. Meaning in science can only be of the second kind.

§77 Meaning for science

Science assigns no conscious intentions to nature. We may certainly ask about the meaning of the genetic code, even if the DNA has no conscious intent. We may ask about the meaning of an exploding star, even if the star has no mind. We may ask about the meaning of the rain, the meaning of a virus, the meaning of evolution. In a generic sense then, we can say that the manifest meaning of any entity is the response to the entity. Irrespective if there be any conscious intention, an entity bears meaning if the entity elicits a response. Thus, the response embodies the meaning of the stimulus.

Because there may be many different types of response to a single stimulus, a single entity may bear different meanings. The protein embodies one meaning of the DNA code, the selective pressure of evolution may be a different meaning of DNA. The energy and matter released from the star embody one meaning of the exploding star, the death of the star is another meaning. The river, the flood and the growing plants and animals manifest very different meanings of the rain. The infection manifests the meaning of the virus to the patient, the virologist thinks of something else. The generation of species and Darwin's ideas are two different meanings of evolution.

Meaning in the realm of science is the fidelity to nature of the scientist's interpretation of appearances. But how can science know whether an interpretation of nature, an experiment, or a scientific theory is meaningful? How can science judge the verity of competing theories?

§78 Validation of meaning

One scientist who has written about the hermeneutics of science is Gunther Stent (see Gunther S. Stent, Hermeneutics and the Analysis of Complex Biological Systems, *Proceedings of the American Philosophical Society*, vol. 130, no. 3, 1986). Stent asserts that truth will always evade us because the "true meaning" hidden in the text of the Bible, for example, can only be the "meaning intended by God". Now science is not privy to the "meaning" of nature "intended by God", so science can never know with certainty that a theory describes an objective truth of nature as she really is (according to the will of the Creator).

Stent enlarges this argument in his paper "Meaning in Art and Science" (in *The Origins of Creativity*, editors K. H. Pfenninger and V. R. Shubik, Oxford University Press, 2001, pp 31-41). The objective of science, says Stent, is to "communicate truth about the world", but "the ideal of an absolutely objective truth is reached only if God also assents to the proposition." Stent, a practicing scientist, asserts that there is an objective truth, and we would know it if only God would reveal it to us. Alas, we have not heard from God on this issue, so the best we can do is hermeneutics, an interpretation of nature. Science can offer only an approximation of the truth, continuously updated by new and more knowledgeable experimentation.

(Plato and other classical Greek thinkers believed that absolute truth was directly perceivable to philosophers—see the parable of the cave in Book VII of *The Republic,* and other dialogues.)

The Talmud, although written about 1,000 years before the advent of Western science, is less naïve regarding our access to absolute truth. The assent of God to the proposition put foreword by Rabbi Eliezer, as you recall, did not free the Rabbis of their responsibility to interpret the meaning of Akhnai's oven [§21]. The Rabbis did hear from God on this issue, but they still had to make their own interpretation. It (the meaning of God's Torah) is not in heaven, says Rabbi Yehoshua, but is a matter for consensus among scholars on earth. Even God's expressed opinion of the *Halakhah* regarding a skin lesion is subject to differing interpretation [§28]. Is there really an objective truth, eternally true, of nature?

The Talmud's most telling analysis of meaning, as we discussed earlier, is to be seen when Moses ascends to heaven to receive the Torah from God, and finds God embellishing the letters of the text with tags and crowns—serifs [§54]. Why, asks Moses, do You bother fiddling with serifs? So that Rabbi Akiva will be able to interpret mountains of *Halakhah*, answers God. Has God no absolute truth to offer Moses and Akiva, only embellishments for interpretation? Moses asks to see Akiva in action. Moses, transported into the future to Akiva's Yeshiva, is satisfied with what he sees, despite the fact that he cannot really understand the Yeshiva proceedings. Moses is happy merely to see that Akiva is making the most of his opportunity to interpret the Torah. Now Moses understands why God bothers providing serifs for interpretation; that's the best thing for humans He can do.

So too does nature afford humans the opportunity for interpretation of nature; that's the best she can do.

(But is there truly no hope of finding absolute truth? Are we to be consigned forever to mere representations, to metaphors? Consider the following: Nature is an infinite continuum, but each species, us included, has evolved to interact only with a particular niche—the limited part of nature that the creature needs to survive. The dog lives in a world of scents beyond the capacity of our olfactory cortex; the bat lives in a world of sound and flight beyond the capacities of our auditory apparatus and our limbs. Who can imagine the world of a blade of grass? A species exists only because it has evolved the biologic machinery for carving out of reality its own place and its own livelihood. True, we humans have open brains, and evolving human culture has learned to see vast scales of nature, from quarks to galaxies. Yet our brains too are limited and will never provide us with more than a skewed glimpse of reality. We will never know more than we can interpret, not even if God were to show us even more than we can interpret. So we are stuck with the use of our own minds.)

Let us, in summary, contrast the position of the Talmud with that of a Scientist (represented hypothetically by Gunther Stent) regarding *meaning* and *truth.* The Scientist (like Plato) assumes that nature has objective meaning, that an absolute truth does exist. Now, if God exists (which in any case is beyond the realm of the scientific method), then He must know true meaning. But there is no scientifically validated revelation by God of that meaning; so scientific experimentation and interpretation is our only option—an option of default—in the approach to ultimate meaning. In other words, there are four elements of interaction: God, revelation, the true meaning of nature and we. True nature and we exist; God and revelation are problematic. So we can never reach the absolute meaning of nature, but only approach such meaning through scientific interpretation. (Plato insists that the philosopher can grasp absolute truth; few now would agree with him.)

The Talmud, in contrast to the Scientist, has no problem with either God or revelation: both, along with us, surely exist. The problematic quadrant is absolute meaning. Meaning, in the view of the Talmud (as I interpret Baba Metzia, 59b [§21], Baba Metzia, 86a [§28], and Menahot, 29b [§54]) is not a static entity, but a process. Interpretation, rather than being a faulty substitute for truth, is intrinsic to the development of meaning. There is no meaning beyond interpretation. Human interpretation, according to Menahot, 29b [§54], is God's truth. So we had better do it right. Doing it right, according to the Talmud, is through dialogue. The right interpretation is reached through open discussion. Right interpretation is a process of consensus. The discussion process, however, is not open to the uninformed public; binding consensus is the prerogative of knowledgeable peers who are recognized experts [§67]. This Talmudic view of right interpretation actually suits the definition of science proposed by the scientist and educator James B. Conant (president of Harvard University from 1933 to 1953). Unlike Stent and his desire for absolute truth, Conant defines science as "an interconnected series of concepts and conceptual schemes that have developed as the result of experimentation and observation and are fruitful for further experimentation and observation" (James Bryant Conant, *Science and Common Sense*. Yale University Press, New Haven, 1951, p25). Science, according to Conant, is a self-perpetuating process: good scientific ideas lead to good experiments that lead to more good ideas and better experiments.

We can summarize the validation of meaning thusly: whether or not one believes in the existence of absolute meaning, science and the Talmud both hold the view that meaning, as we can know it in this world, emerges through discussion, interpretation and consensus within a community of professionals. The rules of admission to the peerage (both scientific and Talmudical) are critical, but beyond the scope of our present discussion.

§79 Biological interpretation of meaning: Co-respondence

Decision-making by collective discussion and consensus is characteristic of human societies at their best—scientific, talmudic, political and familial. The collective interpretation of meaning, however, is not limited to any single scale of reality. It seems that systems of cells, like societies of humans, can also interpret meaning through fruitful consultation.

Consider the immune system. Along with maintenance and repair, the immune system is responsible for protecting our body against invading viruses, bacteria, complex parasites, and whatever else does not belong to us (for more information, you can read a standard immunology textbook). If all potential invaders would act the same, it would be possible to fabricate a standard response to deal with them all; all invaders would mean the same to the immune system. The problem is that each species of invader has a different lifestyle and biology, and so each must be dealt with in a certain way. Some viruses and bacteria are harmless or even helpful, and live peacefully with us for a lifetime (or for as long as the immune system keeps them quiet). Other invaders can endanger our health in amazingly diverse ways. In other words, each invader is best dealt with using a different type of response—different invaders mean different things. The immune system, in short, has to decide the meaning of every virus, bacterium and foreign substance that enters the body. The wrong response might fail to neutralize the pathogen or might even harm the body by producing a debilitating allergy or a lethal autoimmune disease.

The immune system has a large number of possible responses in its armory. The millions of different cells that compose the system each have various response options; each cell can decide to act in different ways, qualitatively and quantitatively, for different amounts of time and in different locations. The responses of a cell or of a few cells, however, are not very significant; the meaning of the immune reaction depends on the collective responses of thousands of different cells. The individual cells have to coordinate their activities to generate a meaningful response at the scale of the organism. Thus, immune decision-making takes place at two scales: at the scale of single cells and at the scale of coordinated cell populations.

A coordinated response would seem to be difficult because each immune cell is autonomous. Immune cells patrol the body individually, and the information available to each cell is restricted to the molecules that the cell momentarily senses with its receptors. An immune cell has no concept of foreign invaders; the cell "knows" only what its receptors see. Moreover, the immune cell's view of the body and the world is only fragmentary; immune receptors see only fragments of the molecules that make up a virus or bacterium. Each cell can "grasp" only a tiny token of the situation.

To complicate the matter, populations of immune cells can be separated into particular families by their structures and behaviors; each family of cells is specialized to sense only a certain type of fragmented information. Thus, the various classes of cells that make up the immune system, for the most part, respond to different molecular worlds. Each cell type gets to see an entirely different and private "sample" of the invader. The information needed to "understand" the problem facing the body is distributed as diverse bits to many different cells. The system has no single command cell; there is no fixed "brain", no discrete "director" directs the action. Pluralism reigns. How can autonomous cells decide what to do individually and then coordinate their activities to make decisions at the scale of the system? How can the poor cells, each with its individualized sample of information, decide how to generate a meaningful response? They do it by a process I have called *co-respondence* (see *Tending Adam's Garden: Evolving the Cognitive Immune Self*, cited above [§69]).

Co-respondence means that different types of immune cells respond together; each immune cell simultaneously receives information about the infection from two sources: At the same time the cell sees its bit of the invader, the cell also sees the effect of the invasion on its fellow cells, which cells themselves are in the process of responding to their fragmented samples of the invader. Co-respondence allows each autonomous cell to respond to its private view of the infection and also to take into account how its fellow cells respond to their own differing views. Each immune cell, therefore, responds by integrating two types of information: invader information and fellow-cell information.

Co-respondence, however, does not abolish pluralism: The individual cells continue to see uniquely individualized pictures of the invasion; no two cells share the exact same worldview. Nevertheless, the cells locally begin to modify their individual responses in the light of the diverse samples of the invasion sensed by the group as a whole. The collective co-response acts to create a high-level picture of the situation, beyond the fragmented worlds of the component cells. Individually, each cell has access to little bits of information, but, through co-respondence, each cell benefits from the diverse experience of its neighbors—the system becomes enriched in information globally. Co-respondence generates, as it were, a systemic abstraction of the situation; no cell alone could ever approximate the complexity of the composite view formed by the interacting cohort of different cell classes.

Note that the global response is dynamic; changes in the responses of a part of the system's cells can spread by co-respondence to generate global changes throughout the system. We outside observers look at the complex behavior of the immune system and imagine that there must be a brain somewhere that understands the meaning of the invasion. No, there is no central command; meaning is generated by the collective process locally. The meaning does not precede the response; the meaning *emerges* as the response proceeds. In fact, we might say, metaphorically, that a locally co-responding cohort of autonomous cells forms an ad hoc "brain". The "brain", which is composed of the interacting cells, persists just as long as the cells continue to interact. When the process terminates, the cells die or disperse to return to patrolling the body. Collectives of co-responding immune cells convene to generate a local "brain" whenever and wherever immune decisions have to be made.

One might think that the autonomy of the cells and their individually unique worldviews would impede the generation of collective meaning. On the contrary, meaning is greatly enriched by the pluralism of the cell population. A composite, higher-order decision can take place only when the individual cells actually see different views of the situation and only when the response of each cell becomes known to other cells. This open "discussion" among autonomous cells is the critical element in immune decision-making because it

amplifies the information available to the system. A co-response is not a mere *averaging* of individual cell responses to the invader; the co-response makes it possible for the cells to "confer" with each other so that the collective of cells—the system—generates a view of the infection that takes into account the diverse views of its autonomous member cells. Thus, the impact of the invader on the collective insight fashions the response to the invader, which in fact is the meaning of the invader. The immune response, indeed, is not to the invader alone, but to the process of the infection in which the invader is only the trigger. Pluralism empowers the system to become more than just the sum of its parts. The immune system, in short, is a complex system and the interpretation of meaning is one of its emergent properties [72]. The immune system, as far as we know, has no brain; so even a brainless system can make interpretations, provided it is sufficiently complex.

(For more about immune meaning see I. R. Cohen, Language, meaning and the immune system, *Israel Journal of Medical Science,* volume 31, pages 36-37, 1995; and H. Atlan and I. R. Cohen, Immune information, self-organization and meaning. *International Journal of Immunology,* volume 10, pages 711-717, 1998.)

§80 Power of pluralism

It may not be too fanciful to apply the lessons of immune co-respondence to people. Each individual, like an autonomous cell, lives in a unique world [§1] and each sees things that others never see. None of us ever really understands the mind of another person; most of the time, we probably don't even understand our own minds. Human society, like the immune system, is pluralistic. But we, like immune cells, do get to see the responses of others to what they see. And we humans too are very sensitive to the responses of others. We are influenced by "public opinion". We are influenced by what others see, even if we never really know what they see. This is co-respondence at the human scale.

One of the slogans of democracy is *e pluribus unum*: one out of many. The power of democracy lies not in the vote, however, but in the process of co-respondence that precedes the vote—the open discussion of the diverse points of view of the individuals comprising the collective. The pluralism of the "free press", which characterizes the Talmudic discourse and the scientific enterprise, empowers political democracy. The autocratic ruler is limited to one static point of view—that of the ruler. The co-responding collective, because of its pluralism, is more knowledgeable, more resourceful and more flexible than is any single individual. Human society is a complex system that generates very interesting emergent properties [§73].

§81 Paradigms

Pluralism, to be workable, needs constraint. Like a species, each discipline of science carves out of reality its own territory for exploration, exploitation and understanding. Inherent to each field of science is an accepted worldview that defines the field's specialized segment of nature. The physicist, chemist, biologist, physician and social scientist each sees a different world and performs approved experiments according to different preconceived rules of procedure and interpretation. The overarching theory that defines the world of interest and the accepted procedures of each discipline is that discipline's *paradigm*. A discipline without a paradigm is not a field of science. A paradigm is a model that tells the scientist what questions to ask and how to find respectable answers. The paradigm supplies the field with a approved set of associations—a kind of high-level metaphor for interpreting the data. If you wish to learn more about scientific paradigms, you can read Thomas Kuhn, *The Structure of Scientific Revolutions*, cited earlier [§3]. Kuhn defines the paradigm as the "universally recognized scientific achievements that for a time provide model problems and solutions to a community of practitioners". A paradigm is an approved framework for scientific interpretation of a piece of nature. One's understanding of causality depends on one's point of view [§72].

Paradigmatic ideas define and control a field of science "for a time" only; the discovery of anomalies—experimental findings that contradict the paradigm—lead ultimately to a scientific revolution: The old paradigm is replaced by a new theory and a new practice. The new paradigm accommodates the new information uncovered by the field about the regularities inherent in its world of study. Science advances by replacing outmoded paradigms. In other words, science does not examine the bare body of nature; science views the body of nature dressed in paradigms.

(Long ago, when I was a research fellow just learning the trade, I did an experiment that seemed to show that the healthy immune system still contained cells that might attack the body itself, as if the body were a foreign invader. I showed the results to the other research fellow who shared the bench with me. Yes, he said, I too have gotten such results. But forget about them, he said, they must be a mistake—the prevalent paradigm taught that such things were impossible. Nevertheless, I continued to work on the question of autoimmunity in health and disease, and do so till this very day. Now, after 30 years, the paradigm regarding autoimmunity is changing; you can see the change in the disputation recorded in an issue of *Seminars in Immunology*, vol. 12, 2000.)

The Talmud too, as we have seen, does not explore the bare words of the written Torah; the Talmud analyzes its own worldview of the Torah, the Torah dressed in paradigms [§18, §50].

Even naïve understanding is shaped by paradigms. I recall the time when my grandmother saw the first moon-landing on television. She had been born in the 19th Century and had witnessed all the 20th Century innovations of human culture. But a man on the moon was difficult to understand. "Well", said grandmother, "if that man is really there on the moon, I would like them to show me a man on the sun." "That's not possible", we explained, "the man would be completely burned in the sun's fire." "I'm not stupid", said grandmother, "they can send him at night."

Grandmother proposed her experiment (send the astronaut to the sun at night) based on the direct and repeatable observation that the sun's fire goes out at night. Grandmother had a limited formal education, but she was, like most humans, a natural at proposing paradigm-driven research.

§82 Tests of truth

Still, science has to make valid interpretations of the data. How does science decide between the validity of contending theories? How does a science adjust its paradigms? The test of validity is understanding, as we defined it above [§68]; a scientific theory stands or falls on a consensus of scientists regarding the clarity of the metaphor [§69], the ability of the theory to successfully make predictions [§70], and the useful applications provided by the theory [§71]. Albert Einstein put it this way:

> The grand aim of all science is to cover the greatest number of empirical facts by logical deduction from the smallest number of hypotheses or axioms.

A valid theory, as a metaphor of nature, should be concise, logical, elegant, and comprehensive in accounting for the facts of nature. Let's imagine a theory that proposes that a certain bacterium is the cause of a disease.

The theory should lead to at least one testable prediction. If the prediction is not fulfilled, the theory falls (see *The Logic of Scientific Discovery*, by Karl R. Popper, Paperback Edition, Routledge Classics, 2002). If the prediction is realized, the theory earns merit. If we infect an experimental animal with the hypothetical bacterium and the animal does not develop the sickness, we may conclude that the theory about the bacterium is wrong. We would have more confidence in the theory if indeed the infected animals would develop the illness.

A theory earns even more merit if the theory can be applied. If eradicating the bacterium cures the disease, we are really impressed with the theoretical connection between the bacterium and the illness. The logic of the validity test is clear and simple. Or is it?

Let's look more closely at the outcomes of the prediction used to test the proposition that the bacterium is the cause of the disease. If the experimentally infected animal does not get the disease, does that disprove the theory? Not really. Suppose the bacterium does kill people, but the strain of laboratory mouse we have used just happens to be resistant or immune to the disease. Suppose the bacteria were inadvertently killed before injection by soap remaining in the test tube. As every scientist knows, an experiment may fail for countless reasons, many trivial and irrelevant to the true question.

Suppose, however, that the mice do get sick; does that prove that the bacterium is the cause? Not really. What if it turns out that the bacterium carries a virus that is the true cause of the disease? Or what if only that strain of mouse gets sick because of some private genetic reason? Or what if the bacterium triggers some other mechanism that is the more immediate cause of the disease? Similarly, our success or failure in treating the infection may or may not materialize for all kinds of possible reasons that could have little to do with eradicating the bacterium. There is always room for interpretation.

(Juvenile, or type 1 diabetes mellitus is caused by an autoimmune reaction in which the immune system destroys by mistake the cells that make insulin. Based on a particular theory, my colleagues and I developed a treatment designed to arrest the immune attack on the insulin-producing cells. The first results of a clinical trial in human patients indicate that our treatment might actually work: see, I. Raz and associates, Beta-cell function in new-onset type 1 diabetes and immunomodulation with a heat-shock protein peptide (DiaPep277): a randomized, double-blind, phase II trial. *The Lancet,* vol. 358, pages 1749-53, 2001. Now suppose that this treatment continues to prove effective, and eventually the treatment becomes licensed for human use; a prediction of our original theory is tried and found true. Would that achievement prove that our original theory is itself true? Not at all. I myself now know that our treatment activates a mechanism much more complex than the one we envisioned when we designed the treatment. But this would not be the first time that a disease might turn out to be cured by a misconception. The complexity of reality is such that many different theories could logically predict the same outcome. There is always room for interpretation. Besides, James B. Conant has already taught us that a good theory need not be true; it need only stimulate us to make a good experiment [§78].)

Consider this: A science advances by updating its paradigms, no doubt about that. But what about the old paradigms that get replaced? Before being superceded by the new paradigm, the old paradigm was accepted as true by the opinion leaders of the field, the theory was taught to students, it was the subject of textbooks. Was the old paradigm less "scientific" than the new paradigm? Did they get the logic wrong? Did they misinterpret the experiments?

Or consider this: Isaac Newton was one of the founders of the science of physics through his development of classical mechanics and classical mathematical tools. Some three hundred years later, Albert Einstein came forth in turn with theories of relativity that viewed nature in a wholly different way than did Newton. Did Einstein invalidate Newton? Quantum mechanics emerged in Einstein's lifetime to view nature in yet a different way. But Einstein refused to accept quantum mechanics. Was Einstein wrong? Was quantum mechanics right? A physicist will tell you that it all depends on the particular way you do your experiment, on your scale of operations; classical mechanics is good for describing certain phenomena, relativity more valid for others, and quantum mechanics operates at its own special scale. So, as we said, different metaphors suit different scales of reality; different metaphors suit different purposes and points of view. Science, like the Talmud, thrives on differing points of view [§67].

What then does science accomplish that is not relative to a point of view? Is science merely interpretation? Clearly not. The data of science are real; data are facts. It's the interpretation of the data that varies with one's point of view. Nevertheless, science uses interpretation to further concrete ends. Science expands and refines its information about nature (the *database*); science develops novel experimental approaches to uncover unexpected regularities of nature; science exploits this information for human benefit (progress); and science expands the limits of the thinkable and the knowable [§78]. The data, methods, applications and ideas developed by science are disseminated, judged, and maintained through communication.

Communication

Texts are central to the scientific enterprise.

§83 Communications of science

Scientists learn their trade and communicate their ideas and observations through the reading, writing and interpretation of texts. The text is central to science, as it is to the Talmud. The point needs no elaboration. A scientific hypothesis is a text; experimentation is translating the text of nature into a representation of nature; a scientific paper is a translation of an experiment into a text for colleagues; peer reviews of scientific papers and funding proposals are textual critiques that bestow approval or disapproval; a scientific textbook details the current paradigm for students using snips of approved experimental texts; science in the media is a text composed for the approval of the public and the government. The philosophy of science is a text written to describe how science is supposed to be done.

Like the Yeshiva [§11], the scientific discourse is an open discussion in which authority can be overruled by a consensus of the majority [§21, §28]. A scientific text is never an infallible proclamation, but always a proposal for discussion and dissent. Science is dialectical; the to-and-fro of public communication is essential. But texts, however much they engage the energy and time of the scientist, come and go. What remains are the data.

Data

Science generates data, enduring information about the facts of nature. But a fact of nature will not become accepted as part of the database unless the fact is noticed, interpreted and communicated by prepared minds.

§84 Database

Scientific data (from the Latin *datum,* that which is *given*) are the observed results of experimentation; the data are considered to be the raw facts of nature *given* to the senses directly and free of interpretation. The collective data are called the database. The database, to my mind, is the keystone of science. The Talmud, as we have discussed in the preceding section, shares basic human values with science; science, like the Talmud, employs interpretation; both enterprises advance through the debate of knowledgeable peers, and both crown consensus. Science, however, differs radically from the Talmud in both the content and the concept of its database. The database of science encompasses the facts of nature and her regularities that have been discovered by scientists and reported in their texts. The database is a flux of information ever enlarged by ongoing empirical experimentation. The texts of science proliferate through interpretation of the data.

The database of the Talmud, in contrast to that of science, is not composed of expanding vistas of nature, but of a fixed primary text, the Torah [§9, §10]. The Torah is interpreted in new ways but is never enlarged by empirical observation to include additional texts, another Torah. True, the Talmud itself has become a database for the *Halakhah* and the Talmud still serves as a database for renewed generations of interpreters. Indeed, the major interpreters of the Talmud like Rashi

Data

and Maimonides have become transformed into databases for following generations. The Talmud and the Torah together serve as a database for the development of the Kabbala and its innovative understanding of the world and God. Judaism grows and thrives like science, but Judaism replaces nature with texts. Its experiments are thought experiments and are never empirical. The texts of Judaism, like those of science, multiply. But Judaism accumulates an enlarging database of thoughts, rather than an enlarging database of facts. The technology that emerges from the Talmud is human behavior; the technology of science, in contrast, has generated the new nature that surrounds us.

§85 Creating the database

Science fashions our understanding of nature out of its accumulating data. We have been led to believe that clever experimentation, precise measurements, confirmed results and the discourse of experts about the data generate clear understanding—if not automatically, then without much interpretation. Understanding increases, according to this belief, as information increases; the more data we get, the wiser we shall be. The problem is that the data themselves may go unrecognized as facts if the scientist is not prepared to understand them, to interpret them as facts. The regularities of nature, the facts, need to be actively mined by prepared minds. Let's take an example from the discovery of penicillin.

The discovery of penicillin led to a revolution both in human health and in our understanding of bacteria, antibiotics, and the host-parasite relationship. Penicillin paved the way for the discovery of the other antibiotics, like streptomycin, that were out there waiting to be noticed. Let me ask a simple pair of questions: who discovered penicillin and when was it discovered? The textbooks will tell you that Alexander Fleming published

his discovery of penicillin in 1929. That seems to answer both questions—it was Fleming and he did it in 1929. But if you look into the matter (read, for example, *Howard Florey. The Making of a Great Scientist*, by Gwyn Macfarlane, Oxford University Press, 1979) you will discover that Joseph Lister had actually discovered penicillin in 1871, and had even used penicillin to cure his own skin infection. Joseph Lister, later Lord Lister, was a great scientist, why did he ignore his discovery of penicillin, one of the great facts of nature? Indeed, why was Fleming's paper of 1929 forgotten until it was discovered by Ernst Chain in 1938? Chain and Howard Florey went on to study penicillin, learned how to produce it on a large scale, and began to supply it for use in the 1940's, during World War II. So we can say that the penicillin revolution started in 1940; was its discovery in 1871 (Lister), 1929 (Fleming), or 1938 (Chain and Florey)? Indeed, we now know that others too had experimented with penicillin in the time between Lister and Fleming. But clearly the *effective* discovery of penicillin was initiated by Chain's discovery of Fleming's paper. But why did it take 67 years, I repeat—67 years, for one of the most momentous facts of nature to become effective, to become meaningful "data" for science? The scientists who saw penicillin but failed to notice it were not stupid; they were among the great minds. The data were there, but they were not understood; they were beyond the pale of the ruling paradigm. It took 67 years for observation of penicillin to enter the database. It takes interpretation for an observation to become scientific data, to become science. It takes hard work for data to be admitted to the database. Data are not to be taken for granted. Data, despite the name, are not so much *givens* as they are *takens*. Data, like the Torah, are given to all but received only by those prepared to see what the data mean [§52].

CREATION

PART IV

Postamble

This book has connected Talmudic texts with ideas related to science. In closing, we consider some agreements and some conflicts between the two systems of thought—religious and scientific.

§86 Definitions: Religion and science

The word *science* is derived from the Latin *scientia*, which means *knowledge*. The origin of the word *religion* is less certain, but seems to derive from the Latin *religare*, which means to *bind*. Science is a way of knowing; religion is system of obligations. Thus religion and science would seem to refer to quite different entities. The Oxford English Dictionary (Second Edition, 1989) emphasizes this difference when it defines religion and science thusly:

Religion:
4. a. A particular system of faith and worship.
5. a. Recognition on the part of man of some higher unseen power as having control of his destiny, and as being entitled to obedience, reverence, and worship; the general mental and moral attitude resulting from this belief, with reference to its effect upon the individual or the community; personal or general acceptance of this feeling as a standard of spiritual and practical life.

Science:
2. a. Knowledge acquired by study; acquaintance with or mastery of any department of learning.
4. a. In a more restricted sense: A branch of study which is concerned either with a connected body of demonstrated truths or with observed facts systematically classified and more or less colligated by being brought under general laws, and which includes trustworthy methods for the discovery of new truth within its own domain.
5. b. In modern use, often treated as synonymous with 'Natural and Physical Science', and thus restricted to those branches of study that relate to the phenomena of the material universe and their laws, sometimes with implied exclusion of pure mathematics.

Thus a religion *binds* a person to particular beliefs about a higher power and obligates the person to worship that power. Science, in contrast, refers to *knowledge* of the material world, and prescribes ways to obtain and classify that knowledge.

Science and religion, according to these definitions, encompass entirely different realms of thought and action, with little or no overlap between them. Since the ascendancy of science, many thinkers have taught that religion and science have nothing in common in their content or mode of interpretation (see, for example, Henri Atlan, *Enlightenment to Enlightenment: Intercritique of Science and Myth*, State University of New York, 1993). Eying each other across the divide, the secularists have conceded that religion may be a good psychological support for those who need it; the religious, for their part, have conceded that science is a useful instrument, but one that ignores (and even demeans) the true place of humankind in the scheme of the universe. Here, I have greatly simplified the gap between religion and science. There are many different religions and their adherents, even those nominally within the same religion, say many different things. Likewise, different branches of the tree of science provide different views of the world to those who ascend the tree. But in general the two realms, even when not at war, are thought to have well-defended borders. The message of this book (if a book that ambles so can presume to transmit a message) is that the borders are not well marked: The problem with the standard definitions of science and religion is that they ignore a significant overlap between these two domains of human thought.

§87 Science in action

The Dictionary does not apprehend real science. The essence of science is a process we call the practice of science; recall how Conant defined science [§78]. Science should not be equated to knowledge regarding "*a connected body of demonstrated truths or...general laws*". The "truths" and "classifications" of science are continuously revised in the light of new experiments and new paradigms [§82, §85]. Scientific knowledge slips and slides. Science is solidly grounded, not on scientific knowledge, but on the experimental method and the discourse of scientists. Science is what scientists do, not what scientists know, or think they know at the moment. To summarize what we said at the outset [§2], science presupposes the right of humans to experiment with nature, to intervene in the workings of nature. That right, however, is not a fact of science; experimentation is not justified by experimentation. The right of humans to intervene in nature is an axiom, an article of faith that precedes science. Science pre-supposes a particular view of the place of humankind in nature. Science in its essence is a form of human behavior based on a particular worldview. The aim of scientific action, through its shifting knowledge of nature, is to improve the human condition—what we call *progress*.

§88 Talmud in action

The Dictionary definition of religion is no less misleading than is the Dictionary definition of science. In none of our Talmudic texts did we have recourse to *destiny, faith, worship, obedience,* or *reverence.* Most of our texts deal with *interpretations* of history, human behavior or suffering. These texts, in their essence, reflect forms of human behavior based on a particular worldview. The aim of Talmudic interpretation is not to advance faith or worship, but to advance the humanity of humans [§42]. From this viewpoint, there is much overlap between science and the religion of the Talmud [§78]. It is true that the practice of science and the practice of the Talmud are quite different affairs [§84]. Nevertheless, each discipline reflects a particular worldview, and each builds its system on collective interpretation. Both science and the Talmud endow humans with a special role in the world— that of making some improvement, what we call *progress*. Of course, there are major differences in what each would consider improvement, just as there are major differences in how each expects to bring about the improvement. Nevertheless, the very concept of progress places science and the Talmud in the same neighborhood [§2, §3, and elsewhere throughout the book].

One can always argue that the Talmud is not really a religious text (it does not fulfill the Dictionary definition of a religion). I can only answer that the Talmud is the basis of Judaism, and Judaism is usually considered a religion; certainly, two genuine religions did grow out of Judaism—Christianity and Islam. One may also remind me that I have selected to discuss only those parts of the Talmud that suit my own point of view [§12]. Other Talmudic texts, which I have not cited (or willfully ignored), do relate to *destiny, faith, worship, obedience,* and *reverence* in a way that would please the Dictionary. The Talmud, as we discussed, does not fear paradox and contradiction [§50]. I invite you to study the Talmud, select citations that please you and reach your own conclusions.

One can also argue the other way and claim that I have violently stretched the boundaries of science to include territory owned by religion.

Scientists, you may claim, presume nothing about the cosmic value of humankind when they experiment with nature; they just do their job. This is exemplified, unfortunately, by those few who experiment with humans or animals without due concern and awe. Yet, I might dare say that science is not the only religion whose fundamental principles are ignored by its adherents.

§89 Creation

But philosophy aside, are not science and the monotheistic religions divided by the concrete issue of creation? The Bible states that God created the world in 6 days; Darwin (and many others) are cited to support the claim that the world evolved. Creationism and Evolution are clearly irreconcilable positions. Or are they? Rashi [§7, §49, §53] in his opening commentary on Genesis 1:1 cites Rabbi Yitzhak (of the Talmud) who argues that the Biblical story of the Creation is not essential to the Covenant; the essence of religion is the *Halakhah* code of behavior. The Torah, as a program for human behavior, logically should have begun with the commandment to observe the Passover holiday [§47], which appears in Exodus 12:2. The Exodus from human slavery towards freedom, not the Creation, is paramount to human religion. Rashi (in the name of Rabbi Yitzhak) concludes that the creation story has been included in the Torah only to teach us that God can rightly intervene in history: He made the world, so He can do what he wants with it. Indeed, we saw above how the Talmud interprets the singular creation of Adam to teach us about human individuality [§14].

Rashi, forever current, points out a number of internal contradictions within the Biblical account and concludes that we must agree that the creation story cannot have been presented to tell us anything about the actual events that gave rise to our world. Biblical commentators like Nahmanides (who lived in Catalonia, 1194-1270) argue with Rashi and claim that the Creation is an essential principle of the faith. Nevertheless, Nahmanides agrees that the creation story told in Genesis must not be understood literally; the Creation requires serious interpretation.

(Some would claim that the concept of Evolution too rests primarily on interpretation—not of scripture, obviously, but of nature and genetics. Evolution is the best scientific idea around, but it is not a controlled experiment.)

§90 Human uniqueness

Whatever one may think about the creative power of evolution, the scientific and the religious views of humankind are surely contradictory: science sees humans as a natural species, and religion sees Adam and all his children as unique creatures [§14]. In which way are humans unique? Again, let us consult Rashi. Genesis 2:7 states that God created Adam out of dust,

> breathed into his nostrils the breath of life, and man became a living soul.

Rashi comments on the expression "a living soul" as follows:

> The animals too were called " a living soul" (earlier in Genesis 1:24), but that of Adam has more life because of the addition of individual cognition and speech.

Note that Rashi uses the word *addition* in referring to *cognition and speech* as characteristics of human life. Rashi's comment implies that Man and the other creatures represent a *continuum* of the manifestations of life. The difference between Man and the other creatures blessed with a "living soul" is cognitive. Humans possess more life

than other creatures because humankind possesses a conscious mind and the ability to communicate by abstract symbols (speech). Humans are unique because of the human mind. Many scientists could live easily with Rashi's interpretation of Man's place in a continuum of nature. Life has emerged through the complex evolution of physical matter into cells and organisms. And the human mind, in turn, has emerged from the complex organization of the material human brain [§73].

However, the continuum of life envisioned by Rashi, in contrast to that proposed by science, is not connected to any process of evolution; Rashi never heard of evolution. Nevertheless, Rashi does interpret the creation story to include the addition of complex attributes to more basic attributes. Has Rashi provided us here with another concrete overlap between science and Talmudic religion? The overlap, if it exists, is not extensive: The Talmud, unlike science, teaches that consciousness and speech require God. Humans may create life, but our golems will usually be speechless [§27].

§91 Einstein and Freud

I have sufficiently (or extravagantly) interpreted (or misinterpreted) Rashi. Let us end with some words about science by two scientists who appeared almost a thousand years after Rashi; one of them revolutionized our views of time, matter and energy and the other our view of the human mind—matter and mind: the polar extremes of reality.

Above, I quoted a statement by Albert Einstein [§82];

> The grand aim of all science is to cover the greatest number of empirical facts by logical deduction from the smallest number of hypotheses or axioms.

This is true enough for physics, and probably for chemistry. In my own field of biology, the complexity of nature is so great that Einstein's statement is, at best, an oversimplification. The one axiom that seems to cover the greatest number of empirical facts in the life sciences is that complex systems need complex explanations.

Sigmund Freud proposed a more complex view of science and reality (S. Freud, *Moses and Monotheism*. Vintage Books, New York, 1939; copyright renewed 1967):

> Our imperative need for cause and effect is satisfied when each process has one demonstrable cause. In reality, outside us this is hardly so; each event seems to be over-determined and turns out to be the effect of several converging causes. Intimidated by the countless complications of events, research takes the part of one chain of events against another, [and] stipulates contrasts that do not exist and that are created merely through tearing apart more comprehensive relations.

Freud, apparently shocked by his own challenge to the worldview of science, adds a footnote:

> I do not mean to say that the world is so complicated that every assertion must hit the truth somewhere.

Note that both Einstein and Freud define science as a process directed at the *interpretation* of nature; like the Talmud, neither of them presumes to reach eternal truth. Is true science then the simplicity of Einstein or the complexity of Freud? Who makes more sense, Einstein or Freud? At this point, I have nothing to add. It's all a matter of interpretation.

About the artwork

The artwork was graciously provided by artist Eleanor Rubin, who creates prints, drawings and watercolors. Her art can also be seen in permanent collections at the Museum of Fine Arts, Boston, MA; the Boston Public Library, Boston, MA; the Institute for Research on Women and Gender, University of Michigan, Ann Arbor, MI; the Alzheimer's Association, Cambridge, MA; and on her websites:

http://www.ellyrubin.com
http://ellyrubinjournal.typepad.com

Eleanor Rubin allowed me to select images that I thought expressed visually some of the themes of the book, give them new titles and insert them where I felt they fit the book. The reader is free to consider each image in the light of his or her own associations and to agree or not with my choices. Interpretation is a free enterprise [§67].

Artwork titles and dates

Person: "Persephone" (watercolor) 2000
Place: "Tuscany" (watercolor) 2009
Time: "Time and Tides" (fabric collage) 2009

Consensus: "Rhythm with Pauses" (watercolor) 2004
Creation: "You Can not Fold a Flood" (watercolor) 2007
Creative Urge: "Venice Sun" (watercolor) 2009
Data: "Floating Possibility" (watercolor) 2002
Hermeneutics: "A Season of Changes" (woodcut) 2003
Meaning: "Attachment and Yearning" (sumi ink) 2003
Progress: "Fugue" (woodcut) 1992
Signification: "Winged Companion" (watercolor) 2002
Texts of Science: "The Story Before She Told it" 2004
Suffering: "A Fearsome Appetite" (etching and chine collé) 1999
Understanding: Untitled (papier decoupe) 2007
Blessing: "Fullness of Being" (watercolor) 2004
The Day Begins in Darkness: "Between then and now" (woodcut) 2003
Atonement: "for Jacqueline Du Pré" (watercolor) 2005
Judgment: "Celebrate Community" (watercolor and papier decoupe) 1992
History: "Progress of the Eclipse" (woodcut on handmade paper) 1999